SCIENCE STORIES

Teachers and Children as Science Learners

Janice Koch Hofstra University

Houghton Mifflin Company   Boston   New York

SENIOR SPONSORING EDITOR: Loretta Wolozin

ASSOCIATE EDITOR: Lisa Mafrici

ASSOCIATE PROJECT EDITOR: Rebecca Bennett

SENIOR PRODUCTION/DESIGN COORDINATOR: Carol Merrigan

SENIOR MANUFACTURING COORDINATOR: Priscilla Abreu

COVER DESIGN: Rebecca Fagan

COVER IMAGE: Copyright Katherine Mahoney

LINE ART: Maria Sas, NuGraphic Design

Acknowledgment is made to the following sources for permission to reprint selections from copyrighted material:

Definition, p. 3: Copyright © 1996 by Houghton Mifflin Company. Reprinted by permission from *The American Heritage Dictionary of the English Language,* 3d edition.

Excerpt, p. 190: Reprinted by permission.

Excerpt, p. 289: Reprinted by permission.

Excerpt, pp. 304–306: Adapted with permission from *National Science Education Standards.* Copyright © 1996 by the National Academy of Sciences. Courtesy of the National Academy Press, Washington, DC.

Checklist, p. 320: "Children's Science Journals: Tools for Teaching, Learning, and Assessing," by Daniel P. Shepardson and Susan J. Britsch, Vol. 34 (5), pp. 12–17. Reprinted with permission from NSTA Publications, copyright © 1997 from *Science and Children,* National Science Teachers Association, 1840 Wilson Blvd., Arlington, VA 22201-3000.

Printed in the U.S.A.

Library of Congress Catalog Number: 97-72504

ISBN: 0-395-70800-1

23456789—DC—02 01 00 99

Science Stories helps build a science-teaching knowledge base.

Science Stories

In memory of Beatrice Deutsch

Brief Contents

Contents

List of Science Topics

Many Science Stories and discussions in this book touch on multiple science content areas. This general guide categorizes the sections according to the topics that receive the most emphasis.

Heat Energy

Electricity

Light

States of Matter: Solids, Liquids, Gases

Preface

For the elementary school teacher, teaching science is a challenging task and sometimes it can even be intimidating. This book is an invitation to elementary school science. Or, to put it another way, the book is like a welcome mat for future teachers, welcoming them to the world of elementary school science and getting them excited about it.

I have written this book so that it can be used as a core text in a science methods course for preservice or inservice elementary school teachers. But its approach and content are forged from my own struggles to find material that would address the school science experience as a natural part of the classroom's daily life. In my experience, many future elementary school teachers are apprehensive about teaching science. They don't believe that they themselves are "scientific" or that they can teach students to think like "scientists." Often, like many others in our society, they have a culturally constructed stereotype of the scientist: a kind of mad genius in a white lab coat who does strange, inexplicable things with test tubes. Unfortunately, many conventional texts do little to counter these preconceptions, and when teachers begin to teach young students, they often convey the same apprehension and the same brand of stereotype. This book attempts to break the cycle by showing teachers that they do indeed have a "scientist within." By locating their own scientific selves, teachers can reawaken the joys and wonder of unraveling the mysteries of the natural world and they can share that experience with their students.

THIS BOOK'S APPROACH TO SCIENCE TEACHING AND LEARNING

Besides its deliberate efforts to awaken the reader's scientific self, this book relies on three other key elements: (1) the use of **Science Stories** derived from classroom experiences; (2) an emphasis on constructivist principles of learning; and (3) frequent connections to the *National Science Education Standards*.

Science Stories: Narrative as a Tool for Science Teaching and Learning

Ten years ago, when I began to teach the elementary science methods course at a major metropolitan university, I started telling stories about my own experiences in doing science with children. I found that the stories helped my students understand both what "science" is and how they could facilitate science experiences in their own classrooms.

At about the same time, I was traveling to various school districts to help elementary school teachers create science experiences in their classrooms. I spent many months visiting schools and modeling science lessons in the elementary grades. As I learned a great deal from the children and their teachers, I realized that their stories had important implications for my teacher education classes. Thus my repertoire of stories naturally expanded.

The use of narrative is not new to teacher education, but it is a fresh approach to the elementary methods course. This book presents much of its theory and its practical advice by way of stories about children and teachers actually doing science together. Rather than imbibing generalities out of context, readers can see what the principles mean in typical classroom situations.

To supplement its narratives, this book uses many strategies to encourage readers to think deeply about these classroom scenarios. The stories themselves are set in a framework of discussion that helps readers see the larger picture. Moreover, following each story or set of stories, an **Expanding Meanings** section focuses on both the teaching principles and the science ideas that can be drawn from the narrative. These sections also offer questions and resources that challenge readers to explore further on their own.

Perhaps the most important thing to be said about this contextualized, narrative approach is that I have seen it work in my own courses. Many of my former students have gone on to become excellent elementary school teachers who inspire children to engage in remarkable science explorations. My experience with these science stories leads me to believe that they are an important teacher education tool. I hope you will enjoy using them in your own classes. And as your students begin their field work, they will create more science stories.

A Constructivist Foundation

This book is deeply rooted in the constructivist approach to learning. Chapter 1 lays out some key tenets of constructivism, providing a solid theoretical base, and later chapters develop and illustrate these important principles. The emphasis throughout is not just on science ideas but on *how* children learn such ideas—how they construct their own meanings. The science stories illustrate many different aspects of this making of meaning, demonstrating how teachers can best mediate the process.

Besides advocating a constructivist approach, this book attempts to model it as well. Readers are continually challenged not to absorb the text passively, but to make their own meaning from it—to think about and extend the concepts and examples it presents. I encourage readers to reflect on their own science experiences as well as their interactions with their students. As the subtitle indicates, the book envisages both teachers and students as learners. It sees science as a lifelong journey of exploration and inquiry. Getting this journey started is perhaps the most important task for science educators.

Meaningful Links to the *National Science Education Standards*

Throughout the text, I refer to various aspects of the *National Science Education Standards* (NSES) and to the instructional practices that they imply. Key points of the NSES are emphasized, including the following:

- the skills of scientific inquiry
- active learning
- extended investigations and problem solving
- collaborative and cooperative learning
- integrating technology fully into teaching and learning
- making connections to students' lived experiences
- using community resources
- responding to and learning from student diversity

....

LEARNING FEATURES OF THIS TEXT ....

In addition to the science stories, the vital theoretical background, and the frequent links to the National Standards, this book provides readers with important tools to help them use the material effectively. These tools include:

Focusing Questions. At the beginning of each chapter, the reader is asked to reflect on a few key questions. These questions are not simply a disguised summary of the chapter content. Rather, they try to make a personal connection to the reader in a way that invites thought about the chapter topics.

Expanding Meanings. As mentioned earlier, the science stories in this book are followed by a section titled "Expanding Meanings." Designed to stimulate reflection, Expanding Meanings sections include four parts:

The Teaching Ideas Behind This Story. A list of the science teaching ideas that are demonstrated by the science story. In this section the reader comes to understand the reasoning behind the techniques modeled by the teacher in the story.

The Science Ideas Behind This Story. An explanation of specific science concepts that the reader should understand in order to grasp the science story fully.

Questions for Further Exploration. Probing questions that set readers thinking about matters that range beyond the specific story.

Resources for Further Exploration. Electronic and print resources that expand on the science ideas and teaching ideas behind the story. In the chapters that do not include science stories, the resources appear at the end of the chapter.

Key Terms. At the end of each chapter is a list of key terms; each term is referenced to the page where it is explained.

Marginal Notes. Short phrases in the page margins reinforce important points throughout the text. Many are deliberately designed to provoke the reader's reflection.

Sections at the front and back of the book also provide significant resources for the reader:

List of Science Topics. As a handy reference, the List of Science Topics organizes science stories and other text sections by their major scientific subject matter.

More Resources for Teachers. As another encouragement for readers to go beyond this book into their own explorations of science and science teaching, this section offers a wealth of useful information.

Glossary. The Glossary provides clear, concise definitions of key scientific and pedagogic concepts.

Finally, for instructors who adopt the text, the accompanying *Instructor's Resource Manual* offers my own ideas for teaching a course with this material. It will help you plan ways to use the book and related resources to excite and challenge elementary school teachers and set them on the path to further exploration. Also, the Teacher Education Station Web Site (found at **http://www.hmco.com/college,** then click on "Education") provides additional pedagogic support and resources for beginning and experienced professionals in education, including the unique "Project-Based Learning Space." For more details on all that this web site offers, see the site map on the back inside cover of this book.

ACKNOWLEDGMENTS

The "seeds" for this book were planted many years ago when I began working with Peggy McIntosh, co-director of the National SEED (Seeking Educational Equity and Diversity) Project, a faculty development initiative on inclusive curriculum. At Peggy's urging, I began to tell my science stories with the remarkable precollege teachers of the SEED project. Moreover, through my association with my SEED colleagues, Judy Logan, author of *Teaching Stories* (1997), and Cathy Nelson, co-director of the Minnesota SEED Project, I learned the value of telling our stories and the promise that these stories hold for teachers. I thank them with all my heart for their vision, their friendship, and their wonderful, authentic work.

Clearly, there would be no stories to tell without the teachers from many different areas who invited me into their classrooms and shared their students with me. I want to express my thanks to the teachers from Deer Isle, Maine; from the Unquowa School in Fairfield, Connecticut; from Somerville Elementary School in Ridgewood, New Jersey; from the districts of Port Chester, Tarrytown, and Somers in New York State; and from Beth Am Day School in South Miami, Florida. A special thank-you to my local Long Island schools, where I have visited many times and observed many wonderful science experiences. My special thanks also to the teachers of the Munsey Park School in Manhasset, who let me use their new science room and allowed me to work with engaging, interesting students for a whole semester. Thana Giradhar from Ridgewood Public Schools and Heather Stoller, Lorraine Amdur, Ilana Johnston, and Marie Ciota from Somers Public Schools are well represented in this textbook. Joanne Daly's web site suggestions were also very helpful. Many thanks to Mary Quinn right here on Long Island who generously supplied wonderful duck stories and photos for this book.

My Hofstra University students—both preservice and inservice teachers—are another inspiration for my science stories. With their candor and their insight, they helped me to understand that they needed a new kind of book. They also showed me that stories work! My Hofstra mentors, Maureen Miletta and the late Selma Greenberg, read portions of the text and urged me to complete it. My graduate assistants, Janine Boscia, Gail Sussman, Roe Balian, and Elizabeth Parr, made it possible for me to do seventeen things at the same time.

Although it is possible for me to remember where the ideas for the text originated, it would have been impossible to turn them into a science methods textbook without the vision of Loretta Wolozin, senior sponsoring editor at Houghton Mifflin. She traveled to Long Island one day to convince me that these stories could make a textbook that would speak to the preservice elementary teacher in a way no other science methods textbook could. Loretta was convincing, and I will always be grateful for her persistence, support, and vision. To help me along, she solicited the services of a creative developmental editor, Doug Gordon, who understood from the outset that *Science Stories* was different and important. It would never have become a book without Doug Gordon.

My sincere appreciation also goes to Lisa Mafrici who was there from the beginning when Loretta signed me on at Houghton Mifflin, and to the talented project editor, Becky Bennett, whose production skills helped me handle the countless tedious tasks of creating a usable, reader-friendly text. Several reviewers also offered important guidance and feedback. Sincere thanks in particular to Pamela Fraser-Abder, New York University; Samuel Hausfather, Berry College, Georgia; David M. Moss, University of New Hampshire; Deborah J. Tippins, University of Georgia; and Robert E. Yager, University of Iowa.

At the very beginning, it was my mother, Beatrice Deutsch, who encouraged my own scientific self to develop and flourish. Today, it is my husband, Bob Koch, and my children, Robin and Brian Tarantino and Betsy Koch, who nurture my present-day scientific self with their own appreciation of the natural world.

Janice Koch

Science Stories

1 An Invitation to Elementary School Science

FOCUSING QUESTIONS

- When you were a child, did you wonder about how things in nature worked? Did you ever try to find out by exploring the world around you?

- If you hear that someone is a "scientist," what does that suggest to you?

- How do you think people learn to understand scientific concepts? And why do so many people have difficulty with such concepts?

Thinking back to your childhood, you may remember wondering about the world around you—wondering what things were and how they worked. I remember being puzzled, for instance, about how people could fit into airplanes, because the planes looked like tiny birds in the sky. I also wondered why we never found a tiny chick when we cracked open chicken eggs. Perhaps you wondered why the sky appears blue, why leaves change colors in some regions in the fall, or why cracks in the pavement appear after an icy winter.

If you and I had such thoughts about the world around us, it stands to reason that other children must have had them as well. In fact, I believe there is a childhood scientist in each of us waiting to be awakened. Ever curious about their surroundings, children have an instinct to explore, take apart, and experiment with the things around them. In this way, they propel themselves toward their own important discoveries.

Too often, unfortunately, these early instincts are buried by later experiences, including what can feel like the chore of learning science in school. You may have some vivid memories of sitting in a classroom and being bored or stifled by science. If you're lucky, you'll also remember one or more science classrooms that inspired you to think and investigate.

I firmly believe that it is the task of elementary school science to nurture children's instincts of exploration. If you plan to teach, this book is designed to help you discover and develop the budding scientist in each of your students. It will help you rethink your ideas about science and science learning. You'll see that all the "why" and "how" questions that young children have about their world can wend their way into our classrooms and enrich our teaching of science. And I hope you'll come to understand that science can be a highly personal and engaging experience, for students and teacher alike.

This book has a lot to say about personal experiences. I want to take you as directly as possible into an elementary school classroom to see how elementary school science involves teachers and children learning together. Before I do, though, there are some key ideas that will help you make the best possible use of this book.

WHAT IS SCIENCE, AND WHY TEACH IT? ....

How do you think of "science"?

What do you think of when you think of "science"? Reactions vary to this question. Many new teachers, making word associations, think "test tubes," "laboratories," or "white coats." There are actually many ways to think about the subject of science. For example, we can think of science as a way to explore nature. Some people think of it as a subject that holds the key to understanding the secrets of the universe. That sounds exciting—and a bit mysterious as well. Many people think about a long list of facts to be memorized—a common notion of school science.

Whatever our associations, we should remember that science is basically an area of knowledge created by people—men and women—who devote much of their energies to exploring some part of nature and trying to make sense of it. In this sense, human societies have a long tradition of scientific exploration. Ever since the dawn of civilization, people have studied nature and tried to understand it.

Science as a Process and as a Set of Ideas

Defining science

To me, the best definition for science lies in the dual nature of the subject. Science can be described as both a *process* and a *set of ideas*. The second part

What Is This Thing Called Science?

Over the years, scientists and many others have written about the distinguishing characteristics of science. There is no one absolute definition of science, but here are a few interesting attempts to describe what makes science different from other ways of studying the world:

science   The observation, identification, description, experimental investigation, and theoretical explanation of phenomena.

—*American Heritage Dictionary of the English Language* (1992)

Contrary to popular belief, scientists are not detached observers of nature and the facts they discover are not simply inherent in the natural phenomena they observe. Scientists construct facts by constantly making decisions about what they will consider significant, what experiments they should pursue, and how they will describe their observations.

—Ruth Hubbard and Elijah Wald (1993, p. 7)

Science advances, not by the accumulation of new facts, . . . but by the continuous development of new concepts.

—James B. Conant (1966, p. 25)

Science is the process of "finding out." It is the art of interrogating nature, a system of inquiry that requires curiosity, intellectual honesty, skepticism, tolerance for ambiguity, and openness to new ideas and the sharing of knowledge.

—Robertta H. Barba (1998, p. 227)

Science is forming questions about the way things work and trying to answer these questions through experimentation and observation. It is having an open mind and rejoicing when the outcome is a surprise.

—Meryl Rosenblum, elementary school teacher (1996)

Science is the belief in the ignorance of experts.

—Richard P. Feynman (1968, p. 319)

What do these statements have in common? How do they match your own conception of science?

of that definition, the "set of ideas," is probably familiar to you. The mention of science typically conjures up images of biology, chemistry, physics, geology, and earth science—the subjects that commonly fall under the rubric of science in schools.

The notion of science as a process may be harder to grasp. But this part of the definition has important implications for the "science ideas" part of the definition. As they explore nature, scientists go through a series of steps that help them to learn more about their area of study.

The Process   Science as process refers to the ways in which scientists go about their work. They usually begin by exploring the questions about nature that engage them. The questions themselves are not necessarily obscure. Your early wonderings that I spoke about at the beginning of this chapter could in fact lead to a scientific question. This is true of many young children's wonderings.

What is a "scientific" question?

Then what makes a question a *scientific* question? A question is considered scientific when, in order to find the answer, people usually go through these steps:

- Make careful observations.
- Set up an experiment and explore the results.
- Test their ideas through further experimentation.
- Ask others to repeat the experiments.

These steps are what we mean by the process of science. Scientists collect information through careful observation of nature and natural phenomena. Their ideas are explored and tested by setting up experiments (and, sometimes, by trial and error). Their experiments are repeated by others in the search for consistent outcomes. To help make one set of results comparable to another set, different scientists follow similar procedures in their investigations.

This process is what sets science apart from other ways of knowing the world. Using the process involves a particular set of skills that are often called **process skills.** Observing is an example of a process skill. So are predicting and making inferences, classifying, and planning an experiment, to name a few.

We use process skills every day.

It's important to realize that the process skills come into play not just in science labs, but in all areas of life in this increasingly complex world. Often we use science process skills without even "coding" them as science. Every day, for example, you make observations about the weather and decide what to wear on the basis of your observations. In doing this, you are using two of the basic process skills. In later chapters of this book, especially Chapter 5, we will explore these skills in detail.

A Set of Ideas   Through the process of observation, experimentation and exploration, and repetition, scientists gather concepts about the natural world. These concepts are referred to in this book as "science ideas." To put it simply, they are understandings about the natural world. Most high school science textbooks are filled with these understandings.

What are science ideas?

Science ideas include definitions and explanations of natural phenomena. For example, in science, energy is defined as the ability to do work. Science books define work as the process of a force moving a given mass through a distance. Energy, therefore, is anything that can make matter move. Scientists calculate the amounts of energy required to do different kinds of work in nature. As this example shows, the language of science sometimes uses terms differently from the way they are used in everyday life.

You can see that science's set of ideas is inseparable from science as process. Without the process, there could be no ideas. Without ideas to spur more questions, there would be no need for further processes.

The Value of Teaching Science

Lurking in the back of your mind may be a question about why we teach science in elementary school. These days especially, when there is so much controversy about the problems of teaching basic subjects like reading and math, why do we bother with science? The answer has two parts, corresponding to our two-part definition of science.

What's the point of teaching science to young children?

First, the set of ideas that we are calling science ideas is increasingly vital in our technologically oriented society. Many educators, cultural commentators, and leaders of industry have said that knowing what science is about is crucial for everyone. A large proportion of today's jobs require some scientific knowledge, and even our everyday decisions—from what to eat to which vehicle to buy—can be affected by our scientific understanding or our lack of it. In today's world, it's never too early to begin learning science, and elementary schools are expected to do their part. A good elementary school science program teaches fundamental science ideas and prepares students for middle school and high school.

Equally important, especially at the elementary school level, are the skills that children acquire by engaging in the scientific process. Remember what we said about the process skills' being useful in many fields besides science. In fact, studies show that science in the early grades has the potential to help children become critical thinkers, able to reason carefully, solve problems, and make informed decisions (AAAS, 1993). Elementary school students who engage in scientific activities in their classroom build confidence in themselves as thinkers. By developing skills and confidence, elementary school science prepares children not just for biology and chemistry, but for English, history, and social studies as well—and, in the most fundamental sense, for life.

In later chapters of this book, you will learn more about science skills and their all-important relationship with science content. As you journey through the book, you will also notice more justifications for teaching science, and you may discover further ways to define science itself. Explore these multiple uses and meanings of science. All of them can help us understand what we do with children in the elementary school classroom.

HOW DO STUDENTS LEARN SCIENCE?

Learning about learning is one of the tasks associated with becoming a teacher. In particular, understanding how children learn science is crucial to becoming an effective science teacher. This intricate and complex subject is one of the most exciting challenges in teaching. As you prepare for your chosen profession, you have the opportunity to learn about how children learn not just from books and lectures, but directly from and with their classmates.

First, though, it will help you to have some grounding in the thoughts of other people who have worked with children. Many individuals have made important contributions to our understanding of how children learn. The scholars, philosophers, and learning theorists who have most influenced this book are John Dewey, Jean Piaget, Jerome Bruner, Lev Vygotsky, and Ernst von Glaserfeld. Although these thinkers did not always agree completely, I have developed my own model of teaching elementary science based on much of their thinking. This model is particularly influenced by a family of theories about knowledge and learning called constructivism.

Major influences on this book

Key Tenets of Constructivist Theory

Basic to **constructivism** is the idea that all knowledge is *constructed.* This means that knowledge is not passively received. Rather, it is actively constructed by the learner as he or she comes to experience the world.

This sounds like a simple notion, but, in fact, it is not. It describes a complex and recursive process of engaging in experiences, thinking about those experiences, seeing how they fit in with prior constructions, and then, sometimes, formulating new constructions. Figure 1.1 illustrates this pattern in the way our minds work.

Do you *have* a new idea? Or do you *construct* it?

Piaget and Bruner   One of the key contributors to this understanding of the learning process was the Swiss scholar and scientist Jean Piaget. Beginning in the 1920s, Piaget conducted countless interviews and research studies with children of varying ages, and he was able to describe *stages of cognitive development* that the children passed through—that is, specific times in the children's development when different mental structures began to emerge. Although he recognized that children vary, he pegged his stages of development to general age ranges. For instance, he believed that most children are in the sensorimotor stage between eighteen months and two years of age, and the preoperational stage between ages two and seven. He also believed that a child's progress through the stages is biologically determined. Most important for our purposes, Piaget concluded that the growth of knowledge is the result of individual constructions made by the learner. In other words, Piaget decided that knowledge is not passively received but is actively built up by the learner—a process of invention or creation, not reception. As he recognized, this gives tremendous responsibility to the learner.

Piaget: Stages in a child's mental growth

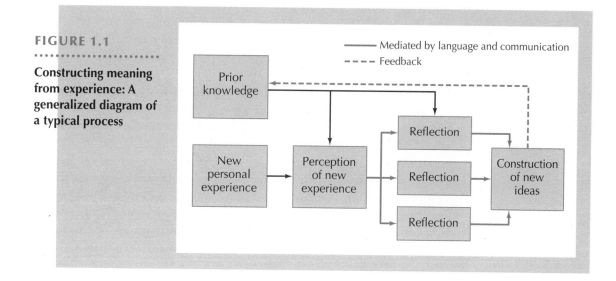

FIGURE 1.1

· ·

Constructing meaning from experience: A generalized diagram of a typical process

TABLE 1.1   Piaget's Stages of Human Cognitive Development

Stage	Age Range	Characteristics
Sensorimotor	18 months to 2 years	Infants explore their environment with their physical senses. They construct such basic concepts as object permanence (the fact that an object can continue to exist even when they do not perceive it).
Preoperational	2–7 years	Children learn language and classify objects into groups, assigning names and categories. They begin to see logical relationships.
Concrete operations	7–11 years	Children build their simple categories into more general categories and apply these in mental operations. Gradually, their thinking evolves from the concrete toward the abstract.
Formal operations	Beginning at 11–15 years	Learners can draw abstract conclusions and master complicated, higher-order processes.

In describing the stages of cognitive development, Piaget demonstrated that at each stage of maturation, the child is ready for a different type of learning. That is, the types of ideas that learners construct vary with each maturational stage. Some critics of Piaget's stages assert that children can be in several stages at the same time and that the stages cannot be neatly defined by specific ages.

Jerome Bruner, a leading supporter of Piaget's work, suggested that at any given stage of cognitive development, teaching should proceed in a way that allows children to discover ideas for themselves. His work became known as **discovery learning.** He differed from Piaget in one important respect. Bruner noticed that children are always ready to learn a concept *at some level,* whereas Piaget believed that readiness for a particular type of learning depended on a child's stage of cognitive development. Realizing this, Bruner emphasized the importance of returning to science topics at various ages, revisiting them at different stages of the child's development (Bruner, 1960). This produces a *spiraling* of science curriculum

Bruner: Discovery learning

topics, as the same broad topics are revisited at higher grade levels. For science topics that are seen as meaningful and relevant for students' lives, this practice has become an important part of curriculum planning.

For many years, Piaget's stages of development and Bruner's discovery learning ideas were the basis for teaching elementary school science. Common to these theorists' views is the understanding that:

- Children are active knowers.

- Children construct their ideas on the basis of their experiences of the world.

- Children's *prior knowledge*—that is, what they have learned from all their previous experiences—plays a crucial role in determining how they integrate a new concept.

Prior knowledge

This last concept, concerning children's prior knowledge, requires further reflection. It demands that we recognize that all students come to the classroom with funds of prior knowledge based on their past experiences. It is common for students to arrive at our doors with varying ranges of experiences and therefore varying types of prior knowledge. As we will see later in this textbook, it is fascinating to learn how students think about new experiences on the basis of where they have been and how they have lived before we met them.

CHILDREN ARE KNOWERS—EVEN BEFORE WE TEACH THEM ANYTHING.

This way of understanding learning requires us to view our students as "knowers"—even before we teach them anything. Indeed, they *are* knowers. Young as they may be, they have lived several years and acquired many experiences before meeting us. Their experiences have led them to construct their own ideas about the world. These ideas become the framework on which they try to fit new ideas. It is as though the children ask themselves, "How does this match my prior thinking? Where can I place it?" Later in this chapter, we will explore how student diversity brings an ever-widening range of prior knowledge to the classroom.

Children learn in a particular environment.

Vygotsky and the Social Context   Piaget's work, which dealt in detail with the individual learner, came under scrutiny for not taking into account the learner's social contexts. Constructing an understanding of the world does not happen in a social or cultural vacuum. The Russian psychologist Lev Vygotsky demonstrated how social contexts influence the ideas that individuals construct as they communicate with each other. For instance, the teacher and children in a classroom use language that is socially and culturally accepted in their specific environment. This language may include slang, regional dialects, or other special usages. The ideas that children construct in the classroom conform to these socially accepted usages and meanings. Brooks and Brooks (1993) put it this way: "Coming to know one's world is a function of caring about one's world. Caring about one's world is fostered by communities of learners involved in trying to answer similar . . . problems" (p. 30).

Implications for Teaching

Although constructivist approaches are not new, they have been drawing a lot of attention recently, and they are producing a major shift in the way educators look at the process of learning. At the same time, a number of researchers, including Noddings (1990) and Brooks and Brooks (1993), have presented ways of looking at constructivism as a methodological tool. By that, I mean that we can translate this theory of knowing and learning into a theory of teaching practice. In fact, this learning theory has helped transform many elementary school classrooms into places where children are active participants in their own learning. That is one of the goals of this book: to help you translate the theory into practice in your own classroom.

Hands-On Experiences   One common classroom technique is the use of concrete activities or experiences for the students. A *concrete experience* involves interactions with objects and materials in the student's real world. These concrete experiences are also called "hands-on experiences" because they typically engage students in exploring materials by physically manipulating them. For example, fourth-grade students studying electricity may work with batteries, bulbs, and wires to construct different types of electrical circuits. Or third graders may take a trip to a museum to explore the remains of dinosaurs for a science unit on fossils and ancient life forms.

Making meaning from concrete experiences

As a result of engaging in such activities, children come up with ideas about the natural world and begin to construct their own ideas and meanings in the way described by Piaget and Bruner. So, when we think of "learning science," we can conceive of it as a process of making meaning from concrete experiences. Also, as Vygotsky pointed out, it stands to reason that this process of meaning making is influenced by children's social interactions. Therefore, when we think about children's learning science, we should imagine their engaging in concrete experiences *along with* their peers and their teachers.

Hands-on activities have been common in elementary school science lessons for decades. However, there are two crucial elements that are often left out:

What does it weigh? Kindergarten children are engaged in weighing objects and recording their mass. Engaging children in inquiry means that they often seek answers to their questions by manipulating materials.

Elizabeth Crews/
Stock, Boston

1. *Connections to the students' own lives.* Because all of us, as learners, try to fit new experiences into what we already know, it follows that we will learn most readily if we can find the proper fit. If children see a school science experience as completely unrelated to their lives, they are not likely to learn from it. To help students see the connections, we need to begin by valuing their own prior experiences and their own thinking. After all, it is through their own thinking that they construct new ideas.

Finding the "fit"

2. *Opportunities for the students to engage in active reflection about their experiences.* This is all important. The process of active engagement must include discussion and reflection in order for the meaning making to happen. When hands-on science became an important elementary science teaching movement, many well-meaning teachers used the experiences but neglected to reflect on them with their students—and not much science learning took place. After all, the real learning is not situated in students' hands but in their minds. Some people therefore prefer to speak of "hands-on/minds-on" science.

Hands-on versus minds-on

Meaningful Experiences   With the basic notions of constructivism in mind, we can say that the most important implication for teaching practice is that the learner must have **meaningful science experiences** in order to construct new ideas about the natural world. A science experience can be considered meaningful if it:

- Relates to the students' everyday lived experiences.
- Engages students in the key processes of science: observing and predicting, inferring and hypothesizing, manipulating objects, investigating, and imagining.
- Stimulates the students to reflect on what they are exploring and come up with their own ideas.

This book can help you construct meaningful experiences in science for your children. It will also help you understand the role of a teacher in this process. Teaching science has a lot to do with providing experiences and listening to the students' ideas. Instead of *telling* students what to think, you will become a *mediator* of their thinking, helping them to learn by reflecting their own ideas back to them and guiding them in sorting out the inconsistencies. You will see many examples of the teacher as mediator later in this textbook.

Telling versus mediating

We can summarize this book's approach to science teaching in the following way:

- Engage children in activities.
- Encourage them to think about the activities and reflect on what they found out.
- Pay attention to the prior knowledge that the children bring to each learning experience.
- Listen to the children's thinking, and engage in conversation regularly about their ideas.

Students' "Misconceptions"   As you can imagine, sometimes children (and adults, too) use their experiences to come up with scientific ideas that are misconceptions—inaccurate ways of understanding the natural world. For example, if you merely observed sunrise and sunset every day, you might think that the sun revolved around the earth, instead of the earth's moving around the sun. Some researchers in science education have called such ideas "prescientific conceptions," while other researchers refer to them as "alternative conceptions."

"Wrong" ideas?

In fact, there is a strong movement away from referring to these ideas as "misconceptions" because that term implies that they are simply

**ALL THE LEARNER'S
IDEAS HAVE VALUE.**

wrong, have no value, and should quickly be discarded. Therefore, *alternative conceptions* is the preferred term. Constructivist philosophy suggests that *all* the learner's ideas—even the ones we think obviously wrong—have some value, since they are part of a process that eventually can lead to a better understanding of the natural world. Again, it is important to value the ideas that learners offer. This is the first step in working with students to help them construct new ways of seeing. Throughout this book, you will find illustrations of how this approach can work in the science classroom.

As the constructivist model of teaching and learning science has evolved, science educators and classroom teachers alike have recognized a need to publish guidelines for educators at all levels. This impulse has fed into the recent national standards movement, a movement that, as you will see in the next section, has an interesting background.

NEW STANDARDS FOR SCIENCE EDUCATION

In the early 1980s, several national reports expressed deep concern about the state of scientific literacy among students at various levels in their schooling, from kindergarten through twelfth grade. *A Nation at Risk* (1983) and *Educating Americans for the 21st Century* (1983) are two documents that warned the American public about the need for all students to have a firm grounding in mathematics, science, and technology. The test scores of American students on international proficiency examinations in science were lagging behind those of students in other countries. Expressing fear that our students were falling behind and that our country would lose its competitive edge, these reports and others sounded an alarm.

Scientific literacy, as defined by these reports, required students to be able to work with the "intellectual tools of the 21st Century" (Educating Americans for the 21st Century, 1983, p. v). Top priority was given to providing earlier, more intensive, and more effective instruction in science, mathematics, and technology in grades K–6. The scientifically literate person was described as an individual who could use the processes of science to make important life decisions. In particular, such a person should have the ability to explore a problem through careful reasoning.

The *Benchmarks*

Part of the response to the alarms of the 1980s has been an attempt to establish **national standards,** first for mathematics, then for science, and, more recently, for technology. One prominent example is *Benchmarks for*

Science Literacy (AAAS, 1993), a document that provides guidelines not only for science, but for technology and mathematics as well. Published by the American Association for the Advancement of Science, *Benchmarks* addresses what students should know and be able to do in science, mathematics, and technology by the end of grades 2, 5, 8, and 12. It explains the value of the processes as well as the content of science. *Benchmarks* is part of a larger movement, Project 2061, named for the year of the return of Halley's comet to earth's orbit. This larger movement involves teachers, teacher educators, and scientists in science education reform.

The NSES

Less than three years after *Benchmarks* came an even more ambitious document. Designed to offer guidelines for teachers, teacher educators, curriculum developers, and school districts, the **National Science Education Standards** (NSES) were published in 1996. These standards make the case that acquiring scientific knowledge, understanding, and abilities should be a central aspect of education, just as science has become a central aspect of our world.

The NSES are divided into separate sections that set forth standards in a variety of related areas, including the teaching of science, professional development of science teachers, assessment in science, and science content. In Part Three in this book, we'll look at the specific areas of science content as we explore how to use these standards to construct an elementary science curriculum. For now, I want to emphasize five themes that are infused throughout the NSES—themes that this book will illustrate repeatedly:

Key themes from the NSES

1. *We learn science by doing science.* This theme relates to the need to engage students in meaningful science experiences.

2. *We learn science by inquiry.* The term *inquiry* refers to the importance of science as process; it implies an understanding of how scientists study the world.

3. *We learn science by collaboration.* This point refers to the importance of students' working with other students and with teachers to share ideas and meanings as they develop their own conclusions.

4. *We learn science over time.* It takes time for students to construct their own understandings from science experiences.

5. *We learn science by developing personal knowledge.* This is another way of saying that students construct their own understandings.

As the last of the five themes makes clear, principles of constructivism are deeply embedded in the National Science Education Standards.

ISSUES OF DIVERSITY

The emphasis that the NSES place on personal knowledge and collaboration should remind us that learning occurs within each person's social and cultural context. In fact, the NSES specifically appeal to teachers to "recognize and respond to student diversity and encourage all students to participate fully in science learning" (p. 32). Today, there are so many different social and cultural contexts within which learners function that it makes sense to find out as much as we can about our students and their personal experiences in order to help them learn science.

Questions to Ask About Your Students

What does it mean for you as a future classroom teacher to "find out" about your students' personal experiences and recognize their diversity? Basically, as we become teachers, we need to ask ourselves several important questions about our students. Wherever you live, whatever the particular types of diversity you encounter, answering these questions will help you become a better teacher:

■ *Who are my students*? This question refers to students' interests, concerns, hobbies, beliefs, and feelings about themselves and others.

■ *What are their lives like?* This question encourages us to explore our students' family and home structure. For example, do they have siblings? Do both parents work outside the home? What do the children do after school? Do they have enough to eat? Who is at home with them in the evening?

■ *Where do they live?* Is their home a room, an apartment, a private house? Is their block or neighborhood safe? Are there friends nearby to play with?

■ *What interactions with nature are possible for them?* How does nature present itself in the neighborhood of your school and community? Are there trees and grass, mountains and plains, rivers and valleys, or is there only concrete and cement where—even so—a dandelion grows?

■ *How do events that shape my students' lives become opportunities to learn science?* In other words, what things happen in your students' daily lives that could be explored through a meaningful science experience? For instance, if your students live near a shore, there are tides that come in and out, and these could be observed as part of a science lesson.

■ *How are my students' beliefs about the nature of science informed by their cultural backgrounds and their gender?* The ways we think about science and scientists are often based on where we live, how we grow up, whether we are male or female, and what cultural group we belong to. In the next chapter, as you explore the socially constructed images of science and scientists, consider how children from different parts of the country, different cultural groups, and different nationalities might construct scientific images.

Connecting Science to All Students

As you can see, learning to teach science involves a lot of learning about the children you will teach. This is increasingly important because diverse classrooms are more and more prevalent in the United States. Global migration has become commonplace, bringing more students from distant countries to our schools. As well, studies have shown that students with special needs, such as those with physical differences and learning disabilities, have far greater chances for academic and social success in the regular elementary classroom than they do in a segregated classroom environment. Consequently, elementary school classes often include students with broad ranges of learning and physical abilities, as well as wide differences in cultural background. Typically, you can expect that your students may differ from one another by country of origin, native language, socioeconomic status, particular learning abilities, ethnicity, and gender, to name just a few characteristics.

Children can make meaning with whatever language and learning abilities they have available to them. As teachers who are engaging diverse groups of students in science experiences, we want to observe how they construct meaning *on their own terms*. It is our job to honor their processes and help them move forward to increase their own understanding.

Helping students make meaning on their own terms

For example, a student who is not a native English speaker may have a complete understanding of the science experiences in which he or she is engaged, but may have difficulty expressing the meaning in appropriate English terms. By encouraging the student in any way possible to communicate his or her meaning, we are honoring that child's process. When we do that, we say, in essence, "Your thinking is valued and appreciated." This understanding prepares the ground for future understanding to emerge in ways that may be more scientifically appropriate.

Constructivist teaching, as noted earlier, has the potential to give children more control over what they are learning. That is, instead of telling students what to think and how to think, we invite them to create their own ideas in ways that are congruent with who they are. For students with physical, emotional, or learning disabilities—who often have been discouraged in traditional science classrooms—this approach is especially vital. And the accommodation of children with special needs is compatible with the methods of constructivism: providing concrete activities, emphasizing the importance of process rather than memorization of facts, promoting cooperative group experiences, and allowing the big ideas to emerge from specific contexts. As you teach, you will learn what accommodations are necessary for your students with special needs. The important point is to abandon the idea that children have to become "normal" in order to contribute to the world—or to learn science.

Students with special needs are not the only population that has not been encouraged in traditional science. For many years, the science teaching methods of recitation and writing have alienated large numbers of students. Girls and people of color have been most noticeably distanced from science; they experienced the subject as unrelated to their lives and disconnected from any contemporary meaning. Studies have shown that

Accommodating students with special needs

SCIENCE IS FOR ALL STUDENTS.

—National Science Education Standards

Issues of access and entitlement are important considerations for technology use. Here, a "girls-only" time at the computer!

Carol Palmer

to encourage the participation of girls and people of color, science teaching needs to be more concrete, make connections to the lived experiences of the student, engage students in social collaboration, and consider topics of contemporary interest (Sanders, Koch, & Urso, 1997; Kahle, 1994; Sadker & Sadker, 1994).

As you can see, these attributes of science teaching that encourage previously discouraged groups of students are the same attributes that make for a constructivist, hands-on/minds-on learning environment. Hence, the national standards—and this textbook—support the idea that good science teaching can encourage the participation of all students. In other words, teaching to encourage diverse groups means good science teaching.

Know your students!

The important point is to *know* your students and engage them in activities that relate to their world. Suppose you had a class with one or more homeless children. You might soon realize that these children, who often go hungry, have a preoccupation with food (Barton, 1997). To engage them in science activities, you might do as one teacher did: choose science experiences that regularly use food, such as making an earth science model with fruits and vegetables (see *Science Experiments You Can Eat* by Vicki Cobb).

As another example, consider a group of girls in one fourth-grade class who thought science was just for boys (Sanders, Koch, & Urso, 1997). The teacher invited a young scientist to discuss her work with the class. Then she helped the students celebrate Women's History Month by engaging them in research about American women scientists. Not surprisingly, the girls in the class began to develop a more positive attitude toward science.

For you as a teacher of science, engaging diverse groups of students in meaningful science learning experiences—and watching what happens as they each construct their own meanings from these activities—can be an exciting opportunity. It is reasonable to expect that, depending on their prior experiences, your students will come up with a wide range of ideas. Hence, the challenge is to mediate the differences and lead your students toward some accurate understandings of nature. This task, not a simple one, can be exhilarating for you as well as your students.

THE ROLE OF TECHNOLOGY

When we think about making connections to our students, we need to explore the ever-changing role of technology in their lives. Because technology is advancing so fast, your students' involvement with it may differ a great deal from your own experiences in elementary school. For instance, even if you spent a lot of time with a computer, it may be difficult for you to imagine how much some of your fifth graders are affected by email or interactive games on the World Wide Web. And if you're not a computer-driven person, some of these activities may be complete mysteries to you. You may need to spend some time experimenting with the types of computer experiences that are commonplace to your students.

No doubt you have been using computers for word processing and perhaps for electronic communication. As you prepare to become a teacher, you will also discover how to use a computer to facilitate your own and the children's understanding of science. Meaningful science experiences can often include the use of computer software, email packages, and research on the Internet. Learning about what is available will be an important task for you—and a continuing task, because new material is created every day. Later chapters of this book, especially Chapter 13, offer some illustrations and guidance in incorporating technology into the science classroom.

Technology in and of itself, however, cannot assist children in learning science. It is the children's *need* to use the computer that makes it part of the science experience. Suppose a fifth-grade class was exploring the causes of different weather conditions. To research the effect of low-pressure systems, the students might turn to the World Wide Web to gather some weather reports. In this case, the students would be using the computer to collect data, part of the process of doing science.

The key point here is that the use of technology derives from a need to know more. Students turn to the computer as part of actively constructing their own knowledge. The technology is not foisted on the student for its own sake; rather, it acts as a needed and useful resource.

Ways of Using Technology

When you ask yourself the question, "How can computers and other technology help engage my students in meaningful science activities?" you will find many answers, depending on the situation. Let's explore a few of the basic categories of technology use in today's science classrooms.

A Resource for Collecting Data   As shown in our example of weather research, computer technology can serve as an important resource for collecting needed data. The World Wide Web has become a major source of science information. Moreover, classroom equipment may come with a computer connection. In one fifth-grade class I visited recently, students were investigating the rate at which a glass of hot water cooled after an ice cube was placed in it. They used a heat probe to measure the temperature every two minutes. The data were automatically transmitted from the probe to the computer screen, where a software package charted them in the form of a line graph. You can imagine how much time was saved by collecting data in this way—time that allowed the students to give additional attention to analysis and reflection on the data.

A computer-linked heat probe

Interactive Problem Solving and Simulations   Some software packages are designed to pose a problem and engage students in interactive manipulation of a simulated natural event. For example, some elementary science software programs allow students to set up a simulated experiment and draw conclusions from their observations. Thus, the students have the opportunity to engage in science processes by manipulating the keys on a computer keyboard. One software program, on the topic of sound, allows students to vary the length, tension, and thickness of a simulated violin string in order to explore the effects these three properties have on the pitch of the sound created when the string is plucked. Using this program, third-grade students have the opportunity to manipulate the three variables in any way they choose. Today, many professional scientists study natural phenomena through computer simulations. In Part Three of this book, we will explore the criteria for using this type of software program effectively with elementary school students.

Simulating a violin sound

Tools for Expression   As we encourage students to write about their ideas and observations in science, computers also become useful tools for them to use in expressing themselves. As you have probably discovered, ideas and draft reports can be saved, changed, and refined on computer disks with a minimum of effort. This can be accomplished with standard word processing applications. As well, computer technology offers students the capacity to display their understanding of science ideas through multimedia presentations. Along with words, a multimedia software package allows the user to present sounds and images to express a concept. In our example of the simulated violin string, the student might use a multimedia software package to present graphic and audio illustrations of strings producing different types of sounds depending on their length, thickness, and tension. The graphics and sounds would be accompanied by explanatory text created by the student.

Multimedia reports

A Means of Collaboration   The most social part of making meaning in science—collaboration or person-to-person communication—is facilitated by electronic telecommunication. Email and other forms of telecommunication allow classes to share their ideas and brainstorm with other students in neighboring schools or across the world. Scientists themselves collaborate in this way, comparing results of their investigations, so it stands to reason that students will find such collaboration useful and instructive.

Brainstorming by email

Throughout this book, you will find references to web sites and other useful computer sources. Instructional technology tools can also include camcorders, VCRs, video-microscopes, educational television, and overhead projectors. You should be prepared to use whatever is available—and whatever will help your students develop their own understanding of science.

Issues of Technology Access

As you explore technological tools for science learning, remember to ask yourself about your students and their lives. You may teach in a community where computers are commonplace, both at home and at school. Or you may teach in a place where children rarely have access to computers at home. This type of information is important as you plan for meaningful science experiences.

Recognizing the importance of a technologically literate society, a number of local, state, and national funding agencies have awarded grants to help schools in less privileged communities purchase state-of-the-art computers. Frequently, however, the wiring and connection lines necessary for telecommunications have yet to be installed, making it impossible to use the computers for networking, the World Wide Web, and email. In circumstances like this, you may need to take the students to a local library or government facility to access the World Wide Web. The hope, of course, is that easy web access will become a reality for all schools.

Even when children have plenty of opportunities to use technology, there can be important differences among individuals and groups. Many studies have shown, for instance, that elementary school boys are more likely than girls to use computers for their science investigations (Mark, 1992; Nelson & Watson, 1991). This suggests that teachers need to promote equal access and equal computer time. One fifth-grade teacher, observing such a problem in his classroom, decided that part of the class's free time would be "girls-only" time at the computer. He also opened the question of girls and computers to class discussion. In doing so, he discovered some reasons for the girls' computer avoidance. It turned out that, in this particular setting, the boys tended to "hog" the computers and the girls did

Are boys monopolizing computers?

not want to fight for equal time. Having a special girls-only time therefore worked well for this class.

No matter where you live or how much computer technology has pervaded your own life, you will find the technology connections in this book useful. True to the spirit of constructivism, I invite you to explore these connections and come up with your own ideas about using technology for meaningful science in the elementary school.

STRUCTURE OF THIS BOOK

This book is structured in a way that will lead you into the world of teachers and children doing science together. Part One invites you to look inward to explore your own inner feelings about science. In Part Two, we look outward to see how teachers do science with children in their own classrooms. In Part Three, we pull it all together and examine some tools to help you create your own science experiences with children.

Part One: The Scientist Within

Approaching the teaching of elementary school science is a complicated task. Frequently teachers ask me for a sure-fire technique that will enable them to be successful. I am sorry to say that there are none, but there is an important place to begin—with your own personal reflections. Whatever the subject you are teaching, you are always teaching *who you are*! How you feel about science and whether you believe it can make a valuable contribution to your students' lives influences how successful you will be as a teacher of science. It is for this reason that I devote Part One of this book to what I call "The Scientist Within."

You'll always teach who you are

We begin with ourselves, locating the scientific self within each of us. This self may be shy after years of lying dormant. This self may be underdeveloped after years of not being validated. This self may be engaging in scientific activities every day and not even be aware of it. Chapter 2 shows you multiple ways to begin discovering your scientific self.

How shy is your scientific self?

In Chapter 3, we turn to children's scientific selves. That is, we examine the ways in which they can construct their own meanings through concrete scientific experiences. As I noted earlier in this chapter, concrete experience is the ground from which we generate our thoughts, reflect on those thoughts, and eventually make our own meaning. So in Chapter 3 we will begin to see what constructivism might look like in elementary school classrooms. This, in turn, becomes the springboard for Part Two of *Science Stories*.

Part Two: Doing Science with Children: Inquiry in Practice

What are science stories?

Woven throughout this book, but most especially in Part Two, you will find my *science stories*, narratives of experiences that I have had or observed as I worked with in-service and preservice teachers in elementary schools. These stories describe what it looks like when teachers and children engage in scientific activities in the elementary school environment. They form the central core of this book by providing a glimpse of children and teachers doing science together. I have selected these particular stories because they explore a range of elementary science activities in different grade levels in the physical, life, and earth sciences.

From each of these stories, certain ideas emerge. Some of them are "science ideas," that is, concepts about the natural world that comprise part of the body of knowledge known as elementary school science. Others are "teaching ideas"—thoughts about teaching strategies, teaching styles, and different methodologies illustrated by the story. Both of these types of ideas are incorporated in the *Expanding Meanings* section at the end of each story or sequence of stories. Also in the Expanding Meanings sections you'll find Questions for Further Exploration and Resources for Further Exploration. The questions invite you to consider other meanings you may derive from your own interpretation of the story. The Resources include notes about science books and professional development books that can assist you, as well as web sites, children's literature, and software programs, where applicable.

Using the Expanding Meanings sections

Notice that in presenting stories, drawing out some ideas, and giving you vehicles to explore these ideas further, I become the "teacher as mediator" that I described earlier in this chapter. It's up to you to become the active learner, constructing your own meaning.

Part Three: Creating the Science Experience in Your Classroom

In Part Three, we investigate the nuts and bolts of doing science with children. We answer questions like: "How do I plan?" "When and how should I modify my plans?" "What should a science curriculum look like?" "How do I assess the science experience?" "How do I involve the community?" and "How do I know I am doing it right?" We also explore questioning strategies and the use of local and national standards that have been developed to guide science instruction.

Overall, Part Three pulls the ideas from the other parts together by considering all the elements of a meaningful science learning experience and modeling how they can be developed in ways that encourage your own and the children's success. Although all the techniques and strategies discussed may be useful to you, the model you construct will be uniquely your own. Teaching is a highly personal activity. Not only do you have to make your own meaning from many well-meaning suggestions, but you also need to ask yourself frequently, "How am I doing?" In that spirit, the book closes with a section on self-reflection, back where we began, to help you answer the "How am I doing?" question for yourself.

Constructing your personal teaching model

BECOMING AN ELEMENTARY SCHOOL SCIENCE TEACHER

Teaching science may seem daunting at first. Science is often the one area that was neglected in our own early schooling. When you chose your profession, it is doubtful that you pictured yourself planning "science lessons." Please know that you are not alone. It is even shocking to some future teachers when they learn that, indeed, science is an integral part of the elementary school experience.

Becoming a learner yourself—and an explorer

Teaching children gives you an extraordinary opportunity to be a lifelong learner. You will constantly be learning about your students, about yourself, and about new ideas. In particular, doing science with children enables you to become a special kind of explorer and meaning seeker. By that I mean that your own honest observations and experiences of nature are an important part of doing science with children.

Because science is both a method and a set of ideas, it has the potential to help you and your students open many doors. Although this is not a "how-to" book with step-by-step instructions on teaching science, I hope it is a how-to-think book—one that will help you think about doing science with children in ways that bring both you and the students knowledge, pleasure, and confidence.

Key Terms

process skills *(p. 4)*
constructivism *(p. 6)*

discovery learning *(p. 8)*
meaningful science experiences *(p. 12)*
national standards *(p. 13)*
National Science Education Standards (p. 14)

Resources for Further Exploration

Electronic Resources

Teacher's Edition Online: Tools for Teachers. **http://www.teachnet.com/index. html.** A World Wide Web site with tips, news, lesson ideas, and connections to other useful sites. Although not exclusively devoted to science, it offers a number of resources for the elementary science teacher.

Print Resources

AAAS. (1993). *Benchmarks for Science Literacy.* Washington, DC: American Association for the Advancement of Science.

Brooks, J., and Brooks, M. (1993). *In Search of Understanding: The Case for Constructivist Classrooms.* Alexandria, VA: Association for Supervision and Curriculum Development.

Duckworth, E. (1996). *The Having of Wonderful Ideas and Other Essays on Teaching and Learning.* New York: Teachers College Press.

Grabe, M., and Grabe, C. (1998). *Integrating Technology for Meaningful Learning.* 2nd ed. Boston: Houghton Mifflin.

National Research Council. (1996). *National Science Education Standards.* Washington, DC: National Academy Press.

Reynolds, K., and Barba, R. (1996). *Technology for the Teaching and Learning of Science.* Boston: Allyn and Bacon.

Sadker, M., and Sadker, D. (1994). *Failing at Fairness: How America's Schools Cheat Girls.* New York: Charles Scribner's Sons.

Sanders, J., Koch, J., and Urso, J. (1997). *Gender Equity Right from the Start: Instructional Activities for Teacher Educators in Mathematics, Science and Technology.* Hillsdale, NJ: Lawrence Erlbaum Associates.

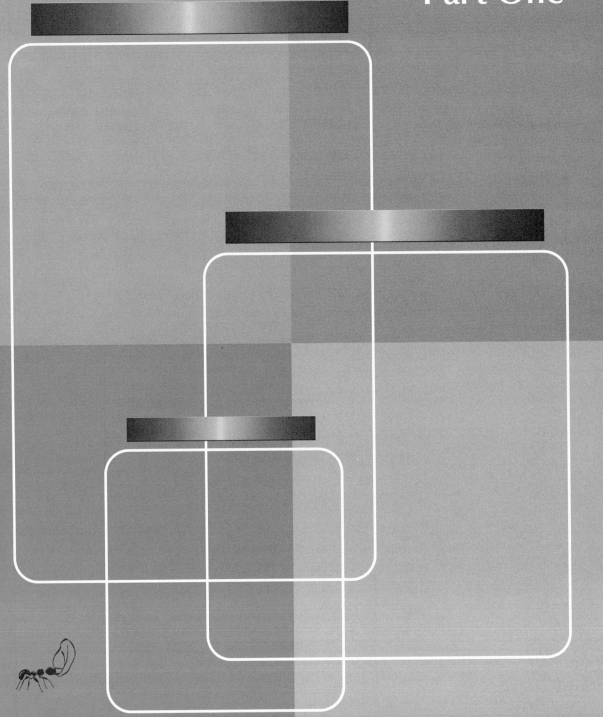

Part One

The Scientist Within

Where I grew up, in an inner-city neighborhood of New York, summertime recreation was usually afforded by the local park. In addition to playgrounds and sprinklers, there were grassy fields and trees. I often occupied myself by lying in the grass and watching the insects on their journey through the grass and clover. Sometimes an ant would crawl up onto my hand and explore my fingers, my palm, and my wrist. In amazement I would watch this ant as it worked feverishly to find familiar ground, knowing it was not on grassy turf.

The park held many other fascinations. In the early fall, I collected acorns from the huge oak trees, as well as "polly noses," the winged seeds from northeastern Norway maple trees. Often I would lie on a blanket, looking up in the sky, and make up stories about the clouds. Their formations became animals or dragons, depending on the day.

On occasions when my family and I visited a restaurant, I would always mix the salt, pepper, sugar, and ketchup into the complimentary glass of water on the table. "Ugh," my father would exclaim, "she's making such a mess." "Don't be silly, dear," remarked my mother. "She is exploring. Maybe she will grow up to be a scientist."

As you can see from this story, my mother, although not herself a scientist, had a feeling about what scientists do, and she "coded" my early explorations as "scientific." You, too, are probably more of a scientist than you realize. Chapter 2 of this book will help you explore your scientific self, especially your personal "science baggage"—the experiences that helped shape your beliefs about science and scientists. You'll write about and reflect on those experiences and start your own science journal. These exercises will help underscore the relationship between doing science with children and feeling scientific about yourself.

In Chapter 3, then, building on your knowledge about your own scientific self, we'll explore the scientific selves of children. Through stories about actual classrooms, we'll investigate in more detail the ways in which students create their own meanings from scientific experiences.

2 Locating Your Scientific Self

FOCUSING QUESTIONS

- How do you normally react to the presence of a scientific expert?
- Do you believe you have scientific skills yourself?
- Why do your personal feelings and attitudes about science matter when you are teaching?

Many new elementary school teachers don't think of themselves as scientists. They often regard science as a subject beyond their grasp. This uneasiness leads them to rely on external authorities in science rather than looking inside themselves and cultivating their own scientific approach. Yet, as I mentioned in Chapter 1, there is a budding scientist in everyone, and the task for a new teacher is to rediscover his or her scientific instincts and develop them. Think about that as you read the following story about an elementary school science workshop for new teachers.

SCIENCE STORY

Why the Balloon Doesn't Pop: An Experience for New Teachers

The science workshop at the state school science conference was titled "Have Fun with Science." In parentheses the announcement said, "New teachers especially welcome." It was a well-attended workshop. There were fifty elementary school teachers gathered in a lab room at the local state college. In front of us stood Rose, an older, gray-haired woman who had bottles of different sizes, colored liquids, yellow balloons, and pins on display at the head lab table.

To begin, Rose pulled two yellow balloons out of a bag and asked us to notice the ways in which they were the same. The teacher audience chimed in with, "Same color, same shape, same size." "They are both deflated," one participant shouted. Someone in the back approached the front table and asked if she could smell the balloons. "Hmm," she mused, "they smell rubbery." Someone in the front shouted, "If you blow them up, they will both be round." Everyone seemed quite engaged with these two ordinary yellow balloons.

How are the two balloons the same?

Rose inflated both balloons so that they were just about the same size. On her work table, she had a small bottle of a blue liquid and a box of straight pins. As she held a straight pin to the side of the first yellow balloon, she remarked, "Of course, we know what will happen when we strike the balloon with this pin." With that remark, she popped the balloon.

Taking the next balloon, she dipped the same pin into her bottle of blue liquid and said, "Now, what do you think will happen if we stick this balloon with the pin that I have dipped in my blue liquid?" The audience was unsure, but the general feeling was that the balloon would pop as the first one did. Rose stuck the blue-liquid-coated pin into the end of the second yellow balloon, and there it sat. With the pin stuck halfway into the balloon, the balloon did not pop!

The mysterious blue liquid—what will it do?

"Now," asked Rose, "why do you suppose *this* yellow balloon did not pop as the first one did?" Several teachers volunteered responses. "The blue liquid," one teacher offered, "sealed the hole the pin was making as it went into the balloon." Another said, "The blue liquid made the pin so slick that it just slid into the balloon without making a hole."

Why didn't the second balloon pop?

"You have made some interesting comments. Any other ideas?" Rose looked around the room.

One teacher, named Susan, sitting toward the back of the room, sheepishly raised her hand. "I thought, as I was watching you," Susan remarked, "that you placed the pin in a different part of the second yellow balloon. You placed the pin in the end of the balloon this time, but the first time you placed it in the side of the balloon."

"What about the blue liquid?" Rose asked.

"Well," Susan responded, "I don't think it has anything to do with it."

Before Rose could comment, I remember thinking how observant Susan had been. I was impressed with her ability to offer an explanation that was so clearly unrelated to the blue liquid.

Rose asked, "How can we test this possibility—that it was where I placed the pin in the second balloon and that it had nothing to do with the blue liquid?"

One teacher suggested that she take a pin, dip it in the blue liquid, and place it in the side of the balloon. Another participant suggested that she take a pin and, without dipping it in the blue liquid, place it in the end of the balloon.

Responding to these ideas, Rose inflated two more yellow balloons, identical to the first set of balloons. She placed a pin in the blue liquid and into the side of the first yellow balloon. *It popped!* Then she placed a pin without any blue liquid into the end of the second yellow balloon. It did *not* pop! Everyone sat back, apparently surprised yet again. Everyone, that is, except Susan.

> Susan is observant.

> What do the tests suggest?

TEACHERS AS SCIENTISTS

Let's think about how the teachers in Rose's workshop behaved. Most of them were ordinary classroom teachers, and some were new to the profession. They were not scientific experts by any means. Nevertheless, they were behaving like scientists in a number of ways:

Observations

1. They made many observations of the yellow balloons. *Observations* include all our perceptions of an object or an event, using as many senses as we can.

Inferences

2. Next, the teachers offered possible explanations for why the second balloon did not pop. These possible explanations are called *inferences*. Infer-

A female research scientist doing field work with mosquito larvae. Challenges in natural science can create a stimulating career for anyone interested in pursuing answers to research questions.

Peter Menzel/Stock, Boston

ences are based on our observations and sometimes lead to setting up further investigations. Each time the teachers came up with their own ideas and thought about the reasons for the events, they were behaving like scientists.

3. While most of the teachers in the workshop were openly expressing their observations and inferences, one teacher expressed a contradiction—a challenge—to Rose, the person directing the workshop and the apparent "scientific authority" in the room. Rose made it appear that the blue liquid was the reason the balloon did not pop. In a way, she was tricking the teachers. Susan challenged this idea and by doing so came up with the correct analysis of the situation—that it was where the pin was placed, not the blue liquid, that made the difference.

Susan's daring is especially important here. Many times, scientific investigations seem beyond our grasp, and for that reason scientific authority goes unchallenged. But once we begin to rely on our own perceptions, we may find that the science in question isn't as obscure as it first seemed. The balloon story demonstrates how important it is to trust your own observations and inferences in order to find out more about the world. In this way, you will get in touch with your scientific self and become *your own source* of scientific expertise.

Challenging authority

Framing a hypothesis

4. By suggesting that the position of the pin had something to do with the balloon's not popping, Susan was creating a *hypothesis*. When an inference becomes a testable idea, it is called a hypothesis. Framed in more precise terms, Susan's hypothesis might sound like this: "The balloon will not pop when the pin is placed in a special part of the balloon on the top."

Testing the hypothesis

5. After Susan offered the hypothesis, the teachers took the next step: testing it. Scientists extend their thinking beyond observations and inferences by reflecting on their ideas and determining how to test them. Rose, following the teachers' suggestions, tested Susan's hypothesis by planning a specific test procedure and carrying it out. These steps distinguish science from other forms of creative human endeavor. Testing ideas by planning experiments and carrying out extended investigations is a critical part of the work of scientists.

Overall, then, the teachers in this workshop engaged in a variety of scientific activities. By observing, inferring, forming a hypothesis, testing it, and, most of all, *trusting their own judgment,* they behaved very much like scientists. Ultimately they came to understand that the balloon is thickest at its rounded end and therefore has the least tension there. A pin placed there will not pop the balloon. If you're doubtful, try the experiment yourself. Make your own observations and trust your judgment.

In the next section, we'll take the next step in locating your scientific self by exploring in more detail the kinds of ideas that teachers, and other people in our society, commonly hold about scientists. As you read, remember the story about the pin and balloon and the mysterious blue liquid.

The secret revealed

By the way, Rose finally admitted that the blue liquid—did you guess?—was just blue food coloring and water.

BELIEFS ABOUT SCIENCE: WE TEACH WHAT WE THINK

You may be wondering, "Why all this fuss about the scientist within and locating your scientific self? If it's just a matter of trusting my own judgment, I can do that."

There is more to it, however. For one thing, looking inward helps you recognize your own feelings about science. Your feelings about the subject of science help to shape your attitudes toward it, and both your feelings

TEACHERS WHO ARE
ENTHUSIASTIC,
INTERESTED, AND WHO
SPEAK OF THE POWER
AND BEAUTY OF
SCIENTIFIC
UNDERSTANDING
INSTILL IN THEIR
STUDENTS SOME OF
THOSE SAME
ATTITUDES.

—National Science
Education Standards

Students recognize
how you feel

and attitudes greatly affect the way you teach science in the classroom. A number of studies have demonstrated this point. One early study indicated that teachers' negative attitudes toward science led to a reluctance to teach science or even an avoidance of teaching science (Perkes, 1975). Additional research in the past twenty years suggests that science attitude scores can be expected to correlate to the behavior of both teachers and children in the science classroom (see, for instance, Shrigley, 1983, 1990; Koballa & Crawley, 1985). In various ways, both overt and subtle, elementary school teachers' attitudes toward science greatly influence their abilities to be effective science teachers. Let's explore some reasons why this is true.

Your Feelings Show   It is almost impossible to mask negative feelings and attitudes about science when you are dealing with your students. Children have an uncanny ability to see below the surface and get to the heart of their teacher's reactions, even when the teacher is a terrific actor and believes he or she is hiding the truth. Children are quick to recognize contrived behavior, so feeling good about science is not something you can fake. It needs to be an authentic reality for you.

Teacher Attitudes Affect Students' Attitudes   Many studies indicate that teachers' feelings and attitudes about science affect their students' feelings and attitudes. For instance, one study that used personal narratives to explore students' attitudes concluded that the scientific experiences that students reported as positive were clearly influenced by teachers whose attitudes toward science were very positive (Koch, 1990). In other words, teachers who have a positive outlook toward science tend to instill that outlook in their students. In fact, interviews with scientists often reveal that they had inspirational teachers whose love of science motivated them to pursue scientific careers (Koch, 1993). Unfortunately, the reverse is also true. Teachers with negative outlooks toward science often inadvertently discourage their students from pursuing scientific interests. As early as 1969, a study found that elementary school teachers' negative attitudes toward science were passed on to the students (Walberg, 1969). Overall, the literature reveals that the role of the teacher is vital in shaping students' attitudes and aptitudes in science (Jones & Levin, 1994).

Teaching who you are

As we noted in Chapter 1, whatever the subject you are teaching, you are always, in a sense, teaching *who you are*. This means that your authentic self is visible and vulnerable in your classroom, and it has a great effect on your students.

The exercises in the rest of this chapter will help you acknowledge your innermost feelings about science. As you engage in these activities, you will learn more about your scientific self. If you were "turned off" to science in secondary school and college, this chapter should help you become more comfortable with it. The important thing is to acknowledge your feelings and consider "making friends" with science as one of the goals of locating your scientific self.

WHAT IS A SCIENTIST? STEREOTYPE VERSUS REALITY

What do you think a scientist is? What does a scientist look like? How does a scientist typically spend the day? To begin discovering your scientific self, let's explore how you conceive of scientists.

Drawing a Scientist

At this point in your reading, you need to close this book. But first get a piece of paper and a pencil and find a firm writing surface. Close the book, and sketch a picture of what you think a scientist might look like. When you have completed the drawing, open the book again to this spot.

What kind of scientist did you draw?

If you are like most other people, your scientist drawing will show a white male with one or more of the following characteristics: wild hair, eyeglasses, a white lab coat, a pocket protector, and some bubbling flasks. (See Figure 2.1.) Studies reveal that both students and teachers frequently draw this popular image of the scientist (Fort & Varney, 1989). Of course, this image is a stereotype because it exaggerates what real scientists look like. The stereotype is reinforced by the scientist images we see in cartoons, movies, magazines, and the popular press. Such stereotypes become part of our belief systems and influence our future behavior. All too often, they limit what we do and think.

Third Graders Draw Scientists   Not long ago, two third-grade teachers in a local elementary school were interested in exploring their students' beliefs about scientists. Distributing crayons and drawing paper, they asked each student to draw a picture of a scientist and describe what the scientist was doing. The thirty-nine students' drawings contained thirty-one men and eight women. Further, of the thirty-one male scientists, twenty-five had beards and messy hairstyles.

FIGURE 2.1   **Responses by two preservice teachers to the assignment "Draw a scientist."**

Reprinted by permission.

Students' ideas of scientists

One boy added a bubble quote for his scientist that said, "I'm crazy." Another third-grade boy described his scientist as follows: "He is inventing a monster. He painted his face green." Still another boy wrote, "My scientist makes all kinds of poisons. He is a weird person." Another caption on the bottom of a drawing said, "Dr. Strangemind," and on the back the student explained, "He does strange things like blow up things and other crazy stuff." Many of the children described their scientists as "blowing things up," "acting crazy" or "goofy," or working with "a lot of potions." (See the typical student drawings in Figure 2.2.)

Think about the stereotype

You can see that most of these third graders, young as they were, had already internalized the stereotyped image of the scientist. To understand why this is important, ask yourself the following questions:

FIGURE 2.2   Responses by three elementary school children to the assignment "Draw a scientist." Compare these to the teachers' drawings in Figure 2.1.

FIGURE 2.2   Continued

- Who is omitted in this stereotype?
- Does the type of person represented in the stereotype reflect the makeup of any elementary school class you have recently seen?
- If the children in a typical classroom were omitted by the stereotype, how would that make them feel about science?

Effects of the stereotype

Stereotypes Can Be Discouraging As you may suspect, the scientist stereotype can discourage individuals who are other than white and male from seeing themselves as truly scientific. Some of the consequences are obvious. For example, substantially fewer females enroll in science courses than do males, beginning in high school and continuing through college (see Figure 2.3). This gender gap persists despite many types of interventions designed to encourage the participation of girls and young women in science. It is a complex issue, but the way in which we conceptualize "who does science" certainly contributes to the problem.

Think about your own drawing

Reflecting on Your Drawing With this discussion in mind, go back and look at your own sketch of a scientist. What exactly did you draw? Who is your scientist? What made you decide on this particular image? Is the

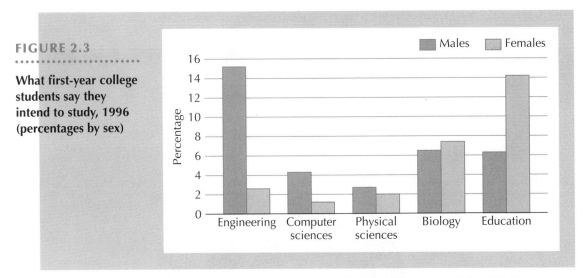

Source: National Center for Education Statistics, *Women in Mathematics and Science* (Washington, D.C.: U.S. Department of Education, July 1997), p. 15, reporting data from Higher Education Research Institute, Graduate School of Education and Information Studies, *The American Freshman: National Norms for Fall 1996* (Los Angeles: University of California, Los Angeles, 1996).

person you drew similar to yourself or very different? What does that tell you about your feelings and attitudes toward science?

Keep the draw-a-scientist activity in mind, not only while you read this book but when you teach science. For both children and adults, the activity can be useful in exploring the implications of the scientist stereotype. You may want to try it with your own students.

Naturally, if people had more contact with real scientists, they would observe much more diversity than the stereotype allows. In the next section, you will be asked to meet with a scientist and ask questions about his or her work.

Interviewing a Scientist

When it seems as if all the images of scientists that you come in contact with conform to the stereotype, it is time to do your own research. Meeting and interviewing a scientist is one way to decide for yourself if the attributes of the dominant scientist stereotype are accurate. There is a problem, however: many of us do not come into contact with scientists as often as we do with other professionals. Depending on where you live, it is not always a simple task to locate a research scientist. The following suggestions can help.

Locating a Research Scientist   Research scientists can be found at private or public colleges and universities, community medical centers, pharmaceutical companies, privately funded research institutions, and government agencies. Your local chamber of commerce can provide the locations of laboratories or corporations likely to employ scientists. You can ask the director of public relations in many of these institutions for help finding a scientist who would be willing to be interviewed.

Preparing for the Interview   When contacting laboratories or other sites where there are research scientists, explain that you are a college student and that you are interested in interviewing a research scientist because you are preparing to teach science in the elementary grades. Explain that you would not need more than one hour for the interview.

When you have the interview date, be prepared; arrive for your meeting with some interview questions and a note pad. The following suggested questions may help you draw up your own list:

Sample questions to ask

What experiences with school science, if any, influenced you to pursue a career in science?

Are there any teachers who stand out in your mind as encouraging you to explore a career in science?

What is the best part of the work you do—the part that gives you the most satisfaction? Conversely, what is the downside of your work?

In what ways is the scientific work you are pursuing similar or dissimilar to the school science experiences you had?

What contemporary scientific issue are you most concerned about?

What would you say to a student who wanted to shape her or his future with a career in science?

If you were to define science, how would you complete this sentence: "Science is . . ."?

You will have your own questions to add to this list, and the interview itself will generate further questions that you could not anticipate in advance.

When teachers conduct interviews of this sort, some of what they learn surprises them. For instance, they learn that many scientists actively pursue family life and religious and political interests. They also learn that sometimes a special teacher was a major influence in an individual's decision to pursue a scientific career. Interviewing a scientist is a way to change the culturally constructed "mad scientist" image that many adults and children hold.

What you learn may surprise you.

Science and Curiosity: An Excerpt from an Interview with a Scientist

Connie (a preservice teacher): What would you do if someone came to you and said, "I hate science, I don't see how it relates to anything, I don't understand it, and I don't see how you can do it"? What would you say to that person to make her or him see that science is a significant field?

Dr. B.: Most people who hate science hate it because they were tortured by it in elementary school. There's a good way and a bad way to teach people science. The bad way is to present science as a set of known things, of facts and relationships that they have to know. Then it becomes a burden. The proper way to teach science is to get the students to understand that science is a way of explaining how the world around them works, and engaging their curiosity. Anyone who has curiosity can't help but find science interesting, if it's presented properly.

Helpful Hints for the Interview   The following hints should be helpful in your interview:

- Do not just ask questions. Engage your scientist in a conversation. Respond to his or her answers with points of your own.

- Many scientists may be eager to hear about your own preparations for becoming a teacher.

Tips for the interview

- Scientists who have young children may share stories of their children's experiences with school science.

- Honor the time frame. Do not exceed your allotted time, unless the scientists freely chooses to do so.

- Some scientists may not be able to talk freely about their research; it may be classified or lead to a patented product. If this is the case, ask them to talk about the general area of research they are engaged in—for example, biomedical research, geological research, theoretical physics research, or something else. You can also ask who funds their research.

- Send a thank-you note!

Becoming more aware of your tacitly held beliefs about scientists is part of the task of locating your own scientific self. The drawing-a-scientist and interviewing-a-scientist activities are two tools to help you reflect on your beliefs about scientists. In the next sections, you are asked to construct other tools to help you locate your scientific self: the science autobiography and the science journal. These exercises can help you explore your beliefs about what science is.

REFLECTIVE PRACTICE I: YOUR SCIENCE AUTOBIOGRAPHY ....

As far back as the turn of the twentieth century, the philosopher John Dewey spoke of the need to educate future teachers in ways of thinking about teaching, not just in the technical aspects of teaching (Dewey, 1904). Since then, the image of the teacher has frequently shifted back and forth between two poles: on the one hand, the teacher as a technical expert, who has acquired and can execute a series of specific skills related to teaching; and on the other hand, the teacher as Dewey's type of *reflective practitioner,* someone who has the capacity to think critically about her or his work. Actually, both facets of teaching are important. Certainly there are particular techniques of teaching and there are ways of thinking that help teachers decide which techniques are preferable in any given teaching situation.

What does it mean to be reflective?

The term *reflection,* when applied to teachers, generally refers to a process in which they explore their own teaching practice, asking themselves important questions about their interactions with students as well as the curriculum. **Reflective teachers** have the capacity to think about their roles as teachers from the standpoint of who they are, who their students are, and what they hope to accomplish. Reflective teachers are very conscious of their feelings about their school environment, their students, and their teaching situation. They are able to recognize potential and existing problems and explore multiple approaches to solutions. To put it another way, a reflective practitioner is always looking inward, asking, "How am I doing?"

Research shows that this type of deliberate, conscious, inward look at teaching and learning can promote your professional development as a teacher. For one thing, it helps you identify areas that you can research and explore. A foremost researcher in the field of reflective practice, Donald Schon (1983), remarks that the reflective teacher constructs her or his practice. What he means by this is that instead of following a list of rules, the reflective teacher explores the classroom environment and creates or modifies her or his own teaching methods to establish new and better contexts for learning.

Since so much of your professional life will involve personal reflection, this is a good time to begin to look inward. In this section, you are asked to reflect on your past science experiences. This type of personal reflection is very important since your ability to be a successful science teacher is influenced by your experiences as a science learner. Whether you know it or not, your own experiences with science in school helped to shape your beliefs about teaching and learning science. Your personal reflections hold the key to uncovering these tacitly held beliefs about science. They also provide the opportunity to begin to modify these beliefs if you feel that is necessary.

Your future depends on your past—and how you deal with it.

Writing Your Science Autobiography

Time to close this book again! First, follow these directions:

1. Take out a fresh piece of paper and a pen or turn on your computer.

2. Think back to your most vivid memories of being a science student, up to and including your college experiences.

3. Are you ready now? You're going to write your own science autobiography.

A personal account

This is a writing exercise in which you explore your feelings about science. You may hand-write your story or use a computer; select the method with which you are most comfortable. Your **science autobiography** should be a *personal* description of your experience with science, in or out of school, through teachers, friends, family, museums, magazines, and other sources. These questions can help you to think about the task:

- When you look back at your science education, what do you see?

- How much science did you study in school?

- Did you like science? Hate it? Did you ever even think about it?

- What personal experiences with school science, scientists, science in the media, and science teachers stand out to you?

Teachers who have written their science autobiographies often begin by saying that they could not remember anything at first. Then they took a few moments to look back and consider their science educations grade by grade, or year by year, and their stories were reclaimed. If you're having

Excerpts from Preservice Teachers' Science Autobiographies

The following passages, drawn from the science autobiographies of five preservice teachers, indicate some of the feelings that elementary teachers have expressed about their experiences with science.

Albert Einstein I am not. Even as a young child, I was not enthusiastic about science. Although I was a very good student and I always excelled in my science classes, including honors chemistry and biology in high school, I was never excited about doing dissections or experiments. When I was given the option in senior year to take advanced placement science courses, or just drop science, I quickly stopped taking science altogether. I have not picked up another science book since.

When I reflect on my past experiences with science, two words come to mind: "Not interested." Then I entered fourth grade, and science came alive. My outlook on science was much brighter, thanks to the efforts of Mrs. M who helped me to feel confident by giving the class activities that we could make our own judgments about.

When I think of science, I can only remember how much I dreaded it. It was boring, rote memorization that never seemed to end. I found it impersonal. Everything that was science came straight from the textbook. You were not expected to understand it, just to know it. I never really understood what I was memorizing, so I think that I never really learned science.

When I teach science in elementary school, it would have to be much different from the way I was taught. It would have to be real—messy, hands-on, something the children could touch, feel, smell—not from a book.

When I was in college, I finally had a good science experience. It was in college biology. For the first time a teacher related the ideas we were learning about to events in the students' experiences. He also allowed the students the freedom to explore their own questions.

Write freely!

trouble getting started, it may help you to read the excerpts above from the science autobiographies of some other preservice teachers.

Write candidly and freely about your evolving experiences with science and your beliefs about it. It does not matter how limited or extensive your experiences are, only that you describe them. There is no prescribed length or format for your science autobiography. It is your story and needs to reflect your experiences. When you have finished writing, go on to the next section.

Reflecting on Your Science Autobiography

REFLECTIVE THOUGHT
INVOLVES "ACTIVE,
PERSISTENT AND
CAREFUL
CONSIDERATION OF
ANY BELIEF . . . IN THE
LIGHT OF THE
GROUNDS THAT
SUPPORT IT."

—*John Dewey*

Think about what
you've written

The very process of writing a science autobiography helps us to examine our remembered experiences. But that is only the beginning. By reflecting on your story, you can come to a deeper understanding of how your experiences provide the background for your current thinking about science and your attitudes toward science.

We know that teachers bring these beliefs—formed by their direct experiences with science in school, the people they meet who work in science, and the publicity science receives—to their teaching practice. Chances are that you will too. The challenge is to reflect on what you have written and make a conscious decision about how these remembered science experiences will influence your behavior as a teacher. You can then use your experiences to forge ahead.

Begin your reflection now. You may want to start a new chapter in your science autobiography at this time. Ask yourself, "Why did I remember some stories and not others? How did these stories frame my current feelings about science and the prospect of teaching science?" As one educational researcher who has worked with educational biographies and reflective journals remarks, "I organize my story as I organize my world and it is my story of the past that can tell me where I am and where I am going" (Grumet, 1980, p. 155).

You may already be starting to see science in a different light as you find your scientist within and consider teaching science in elementary school. In the next section, you can try another device for locating your scientific self.

REFLECTIVE PRACTICE II: KEEPING A SCIENCE JOURNAL

At this point in your preparation for teaching, you may be familiar with reflective journals. These are usually personal journals, kept in a separate notebook, that record many of the feelings and events that have engaged you in your journey toward becoming an elementary school teacher.

Personal journals are nonthreatening vehicles that help you to explore your ideas and feelings about teaching and perhaps about classrooms in which you have been an observer or a participant. Teachers use personal journals as a way to reflect on their practice by writing about classroom situations that require careful consideration. Personal journals are important

Using a journal for
reflection

for developing your skills at looking inward. The activity of writing in your journal and reflecting on what you have written becomes your own process of meaning making, and this is an important tool for your professional development.

What Is a Science Journal?   A **science journal,** as we use the term here, is a personal journal in which the focus of your attention is nature and natural events in your daily experiences. Science can be thought of simply as a way of knowing your natural world. Therefore, a science journal contains your observations of and questions about nature. Often, as we charge through life in this fast-paced world, we are oblivious to the ways in which nature presents itself to us. Keeping a science journal forces us to take more notice of nature. For this reason, science journals are a way of contacting your scientific self.

Science journals can also contain items about science in the news or science that you see in a classroom. You can write about a science show you saw on television. Perhaps you have observed science in a school. Write about it. What did you think of it? Often there are items on television news or in the print media that relate to a scientific breakthrough or recent medical discovery. Write about them.

How Do I Keep a Science Journal?   Keep your science journal in a separate notebook, and make entries in it on a fairly regular basis—about two or three times a week.

Remember that your science journal can encourage you to ask your own questions. If you observe something that you do not understand, write about it. A science journal may contain your mental wanderings and scientific wonderings. You may also think about a science journal as a log, recording the natural events that capture your curiosity and take you by surprise.

A Bird Story: Sample Entries from a Science Journal

The following excerpts from my own journal illustrate a type of story that sometimes emerges in science journals. This is the almost-daily log of a bird-watching experience I had in my own backyard. The experience took me by surprise, and I enjoyed writing about it. The geographic setting is a heavily treed suburban area outside New York City. Notice that I did not know how to explain everything I saw; I recorded my questions as well as my observations.

Your own observations and questions

YOU CAN BE TAKEN BY SURPRISE EVEN IN AREAS WITH WHICH YOU ARE FAMILIAR. IT IS THE SURPRISES THAT HOLD THE POTENTIAL FOR FURTHER LEARNING.

—*Eleanor Duckworth and colleagues*

Tuesday: I never noticed the hollow in the middle of the trunk of the low-lying tree in the backyard. Imagine my surprise today when, as I glanced through the dining room window, I saw a bird fly right out of the tree! I wandered over to the tree trunk and peeked inside the hollowed-out area. I counted eight small white oval eggs. Each egg seemed to be about 4 cm long. The baby birds will soon be hatching, I think.

I'm wondering . . .

I wonder why the eggs are white. Wouldn't it be better if they were a dark color, for camouflage?

Tuesday [two weeks later]: Well, it has been two weeks, and the mother bird continues to fly into and out of the tree periodically. Sometimes when I glance into the hole, I can see her just sitting there. Yesterday when I went to pay her my usual visit, she flew out just before I reached the tree. I leaned over to peek at the little eggs, but they were gone! In their place

A mother flicker guarding her hatchlings at the hidden nest in the hollow of a backyard tree. Notice the black marking at her neck and the ever-hungry hatchlings!

Anthony Mercieca/Photo Researchers, Inc.

were several tiny birds, their necks extended upward and their beaks wide open. They made soft, small chirping noises that seemed clearly to say, "Feed me."

Quietly I stared at them, and then, fearing that the mother bird would return at any moment, I walked away. Now I am wondering what type of bird this is. It's not one that I recognize.

Wednesday: This morning I ran out to the hollow in the tree. The sun was shining at just the right angle, into the hole in the tree trunk. The mother bird had just flown away, and when I peeked in, I counted at least eight pink baby birds, eyes closed, with no feathers. They really do look brand new, and fragile. I wondered where the mother bird had gone and how long she would leave her hatchlings alone.

An hour later, I returned for another glance, and as I looked down into the hole, the mother bird looked right back up at me. Was I surprised! I quickly walked away. But I've noticed that she has a pointed beak and brown feathers. When she flies off, I can see white feathers on her tail. She's a largish bird. I must find my bird book and identify her.

Sunday, late at night: I had to leave on Friday for a short trip, and I hated to miss the progress of the hatchlings. I've observed that the mother bird sits on the hatchlings often. Her frequent feeding trips take her into and out of the hollow in the tree many times a day.

Tonight I arrived home at midnight, eager to see how the baby birds were doing. I took a flashlight and went outside. As I approached the tree, I noticed the mother bird perched inside, looking straight out of the hole, which is several inches above the nest. Her clawed feet must have been clinging to the inside of the tree. With my flashlight shining on her, I could see a black marking at her neck, as though she had on a collar. This black necklace seemed most distinctive. I have to get that bird book! I didn't see the hatchlings tonight—I decided to come back inside without disturbing the mother.

Monday: The hatchlings are still there, a little bigger now. And I've found my field guide to the birds. Combing through the pictures, I spotted several birds that seem to look like the mother bird. She's a little like a catbird, a little like a mockingbird with her white feathers under her tail. But then I found her—and her black necklace. She's a northern flicker, I'm sure of it. I was so excited to identify her.

Wednesday: It has been two days now since I last saw the mother bird. I look often for her, making her frequent runs into and out of the nest, and I'm getting worried. Last night I woke up in the middle of the night and

---

Margin notes:

What type of bird?

How long will she leave them?

Exploring at night

Found her!

What happened to the mother?

went out to the tree with my flashlight, but there was no sign of her at her guard post in the hollow of the tree.

Thursday night: This was the third night, and still no mother bird. I made my decision. The eight baby hatchlings would starve if I didn't do something! I prepared a deep, small cardboard box with wood chips and shreds of newspapers. I was nervous as I reached down into the hollow of the tree to lift the baby birds out, one at a time. Oh, my, I thought, they're down much deeper than it appeared. It took almost the entire length of my arm to touch one. Slowly, I grabbed the bottom of one bird, and then another, and after ten minutes, I had lifted all eight hatchlings from their nest. They felt very warm to the touch.

Earlier today, before finally deciding to remove the birds, I had contacted the owner of the local pet store to learn what he recommended for flicker hatchlings. I drove over there to buy some food and a small syringe for feeding them. Tonight when I put them in the box, I was ready. Using warm water, I mixed the food and filled the syringe. They were very hungry. They nestled together in a corner of the box on a table in my garage. I'm going to get up every two hours tonight to feed them.

Friday morning: It's not easy getting up every two hours to feed birds. But I'm glad I did it.

At 8 A.M. I called the local bird sanctuary and asked if I could bring the birds in for care. I got directions for finding the place. Mary, the director, turned out to be kind and caring. She immediately placed a small heating pad under the hatchlings inside their box. Using her fingers, she put solid, moist food down their throats with her fingers, and she showed me how their necks swell as the food goes down. The bulge in the neck shows which one has been fed. "My, what beautiful flickers we have here," she said. "Where is the mother?" I explained to her that I have not seen the mother bird for days. She said that if the mother was still around, I surely would have seen her as I was removing the babies from the tree. She said it was important to feed them often and that the mother bird would have fed them at least every two hours. So I guess my feeding them last night was the right thing to do.

As I left the bird sanctuary, glad to know that the birds were in good hands, Mary invited me back any time to visit with them. When I returned home, I was surprised to notice how lonely I felt after my days of observing the activity in the hollow of the tree. Now the tree is empty.

Friday evening: I've been realizing that the process of observing nature requires an investment of time and leads to an attachment to your objects of study. I really became *attached* to that bird family, and now that they're

A rescue mission

Feeding the baby birds

Loneliness

gone, I miss the activity as well as the birds themselves. I'm even slightly miffed that I won't have to get up every two hours tonight to feed birds!

When you spend time making observations of nature, I suppose, you become invested in that experience. This afternoon I looked around for something else to study, but nothing caught my eye. I still wonder what has happened to the mother bird. Actually, I'm terrified that she'll return to the tree to find her hatchlings gone!

Monday: I returned to the bird sanctuary today to visit the hatchlings. They were still in the box I had prepared for them, and they seemed much bigger. Their feathers are appearing—black feathers emerging around their necks. I fed them with my fingers and chatted with Mary. In a week, she said, she'll place them in an indoor cage, so they can flutter around. Then they'll be removed to an outdoor cage, where they can practice their flying before being released into the wild.

Mary assured me that they are in very good shape and will be fine adult birds before long. I asked about something else that's been on my mind— why the eggs were white rather than colored for camouflage. She explained that since these birds lay eggs in hidden nests, they don't need camouflage.

Before I left the sanctuary, Mary told me that another flicker may choose to lay eggs in the same tree hollow. Funny—my heart soared! I'll welcome the chance to observe another bird family. I'm going to watch this tree (that I barely knew existed a few weeks ago), and I hope nature will give me another opportunity.

Learning from Your Science Journal

As the bird story from my own journal illustrates, making careful observations of something in nature that engages you is one way to become your own authority about a given phenomenon. I never knew about northern flickers before that episode; I couldn't even identify one when I saw it. But after days of observing and writing in my journal, I certainly knew a lot about the birds in my backyard. In this way, the science journal helps bring the voice of scientific authority inside yourself.

It can also be helpful to share your journal with others. As my episode with the flickers unfolded, I was teaching one of my regular teacher education classes, and I took the opportunity to read my journal entries to my students, all of whom were preparing to become teachers. They all became nearly as wrapped up in the events as I was. Each time they saw me, anxious to hear how the birds were progressing, they asked me, "How are the hatchlings doing?"

Becoming emotionally invested

Another question answered

OUR STORIES ARE THE MASKS THROUGH WHICH WE CAN BE SEEN, AND WITH EVERY TELLING WE STOP THE FLOOD AND SWIRL OF THOUGHT SO SOMEONE CAN GET A GLIMPSE OF US AND MAYBE CATCH US IF THEY CAN.

—Madeleine Grumet

Linking personal observation to teaching

Moreover, as I shared my science journal with these preservice teachers, they began to bring in books and articles about birds that they thought would interest me or that piqued their own interest. It is not unusual for observations of nature to lead to some research, and that is what happened with us. We clipped a local newspaper article on bird watching that featured the northern flicker. We learned that our area has a rare-bird-alert hot line that provides a frequently updated, taped report of what rare birds have been seen lately and where. Some class members contacted the local Audubon Society, which sponsors bird walks, lectures, and other opportunities to learn more about local and migratory birds.

By the end of the class, we had amassed a significant amount of information about local birds and northern flickers in particular. We all understood that this process began with my science journal and the entries I shared with the group. Doing research about a science journal topic often happens when we personalize the experience of observing nature and invite others to share in the event.

As further illustration, the following excerpt explains how keeping a science journal helped one new teacher:

One of the things I am learning as I journal is that the more I look for things in nature to write about, the more I can pass on to my students. I have to do a lot of searching for myself in order to really involve my students. I have always appreciated nature, but I am even more attuned to it since I have been keeping this journal.

The greatest thing about observing nature is that I never know what to expect when I look out my window in the morning. Things are always new and changing and wonderful. All it takes is a little encouragement, and a little time, to take in all there is to observe. Hopefully, I will be able to pass that encouragement on to my own students one day.

It is snowing today. The first thing I will ask my students to do is to talk about the snow in their science journals: What are all the things they notice about the snow? What are their questions?

Some Guidelines for Your Own Science Journal

Keeping your own science journal means taking notice of your natural surroundings. It also invites you to develop a keen ear and eye for news stories that involve scientific matters. Anything can be grist for the mill. Keep these reminders handy as you begin your journal:

Helpful hints

- All your observations are important. No observation is silly or too simple.

- All your questions have value; collect them in your journal.
- Note the date and time of all your entries. This information can be helpful later if you want to go back and look for a pattern or connections.
- Entries in your journal can be of any length.
- Watch for interesting science shows on television.
- Make use of any opportunities to find out what other people are thinking about natural events.
- Have fun and write freely.

THE INNER SCIENTIST

The activities in this chapter have all been designed as tools to aid your self-exploration. They will help you locate your scientific self through personal reflection and exploration of science and scientists. From Rose's yellow balloons to your scientist drawing and science autobiography, they all concern the question, "What do *you* think about science, and how can you feel most comfortable with it?"

Opening doors

Remember, your "inner scientist" will ultimately be visible to your students. I hope that this book will help you approach the teaching of elementary school science with pleasure and confidence. Science has the potential to help both you and your students open many doors to thinking about the natural world. In the next chapter, as you continue to develop your own inner scientist, we will begin to explore your students' scientific selves.

Key Terms

reflective teacher *(p. 41)*
science autobiography *(p. 42)*
science journal *(p. 45)*

Resources for Further Exploration

Electronic Resources

Association for Women in Science. **http://www.awis.org.** The World Wide Web site for this national organization has links to many resources. For example, it can help you in finding a scientist to interview. Click on "Resources" and then on "National Chapters" to find a list of women scientists and their addresses. Or, after clicking on "Resources," choose "Mentoring" to find the email address of a contact person who can help connect you with a scientist.

The Scientist. **http://165.123.33.33/index.html.** This web site of the magazine *The Scientist* is a good place to read interviews of scientists.

UT Science Bytes. **http://loki.ur.utk.edu/ut2kids.** Designed for both teachers and students, this web site has fascinating information about the current research interests of some University of Tennessee scientists.

Print Resources

Dewey, J. (1933; reissued 1998). *How We Think: A Restatement of the Relation of Reflective Thinking to the Educative Process.* Boston: Houghton Mifflin.

Fort, D., & Varney, H. (1989). How students see scientists: Mostly male, mostly white and mostly benevolent. *Science and Children,* 26:8–13.

Koch, J. (1990). The science autobiography project. *Science and Children,* 28(3):42–44.

Shepardson, D. P., & Britsch, S. (1997). Children's science journals. *Science and Children,* 35(2):13ff.

Shrigley, R. (1983). The attitude concept and science teaching. *Science Education,* 67(2):425–442.

Songer, N. B., & Linn, M. C. (1991). How do students' views of science influence knowledge integration? *Journal of Research in Science Teaching,* 28(9):761–784.

3 Children's Scientific Selves

FOCUSING QUESTIONS

- As a child, did you ever invent your own theories for why things happened—theories that were useful at the time but didn't necessarily match what the rest of the world believed?

- Can you remember some of these "alternative conceptions" you had as a child? Do they sound silly to you now, or do you see them as interesting, creative, and maybe even logical?

- Why do you think hands-on science is not enough to help children learn science concepts?

Chapter 2 invited you to begin discovering your scientific self. As you did, you probably revisited your childhood wonderings and wanderings. One new teacher, who grew up in an urban area, remembered wondering why it was always cool in the summer in the basement of the apartment building where she and her family lived. When it got too hot, she and her friends would retreat to the basement.

Children have many questions like that, but too often schools do not make room for these questions, and the children grow up, as this teacher did, with their questions unanswered. This happens because these teachers, no doubt well-meaning, approach science teaching and learning from their own perspective only, asking themselves, in effect, "What do I have to *transmit* to my students?" As you will see in this chapter, science education in the elementary school is much more about the exchanges that occur between teachers and students as they explore science together.

While you rediscover your scientific self, you will also notice the ways in which your students express their scientific selves when you engage them in meaningful science activities. As we mentioned in Chapter 1, children construct meanings from science experiences on the basis of who they are, where they have been, and their own prior understandings. In this chapter, you will see how you can help your students construct their own meanings by listening to their ideas and learning how they think.

In Part Two, we explore some of the many types of science activities in which you and your students can engage. At times, you may engage all the children in a similar experience at the same time, or you may involve different groups in different activities centered on a similar theme. Sometimes you may show the students a natural event by way of a demonstration. That is what happens in the following science story.

SCIENCE STORY

The Bottle and the Balloon

I am visiting Ms. Hudson's third-grade class in a suburb of a major northeastern city. The children come from many cultures— European American, Latino, African American, and Asian American—and a wide range of socioeconomic classes. There are twenty-four students in one large classroom. I'm standing up front with the simple apparatus I've brought for this lesson.

The children watch as I place a deflated red balloon over an apparently empty vinegar bottle. It hangs to the side like Santa's hat. I hold the bottle up to the class and ask, "What do you think is in this bottle?"

A deflated balloon, an "empty" bottle

The children respond, "Air!"

I put this thick glass bottle into a pot of just-boiled water, holding the top to keep it upright. "Let's watch what happens here," I say. Slowly, the deflated balloon fills with air, until it stands erect at the top of the bottle.

The children are delighted. With only a little prodding from me, they carefully explain what they observe, and we record it: When we put the bottle in the hot water, the balloon filled up with air.

"Where do you think the air has come from?" I ask. "What are some of your ideas?"

My thinking: At this point I am wondering if the children will grasp that making the bottle hot will heat the air inside the bottle and that the warm air causes

Science is a way of knowing the natural world. Here students explore nature by examining succulent plants on a field trip.

David Young-Wolff/Photo Edit

the balloon to inflate. What actually occurs is that air that is warmed expands. Its particles move faster and become farther apart. This warmed air is now lighter; the colder air above it, which is heavier, moves down, pushing the warm air upward. I do not expect the children to have that level of knowledge, but I do hope they will make some connection between the heated air and the balloon's inflating. Instead, the children give me some creative but quite different explanations.

"The steam from the boiling water went through the glass and inflated the balloon," a boy named Mike offers.

"That is an interesting idea," I answer. "Have you ever seen steam go through glass before?" I'm not making fun of Mike's idea; instead, I'm trying to draw out his prior experiences that may have contributed to it. But after thinking for a moment, he does not remember having seen steam go through glass.

A girl named Jamila agrees with Mike's focus on steam, but has a different idea about the mechanism: "The steam seeped into the balloon because the seal between the balloon and the bottle was not airtight."

I respond, "That sounds as if you have really been thinking about this. Do you think that my placing the bottle in the hot water helped the steam to seep in?" Jamila says, "Yes, it helped steam get in."

The children's ideas

Why does the teacher keep asking for students' ideas?

It's clear to me now that several students think that steam seeped into the balloon. I continue to question them about their thinking.

"Do you think that if we made the seal between the bottle and balloon tighter, we could prevent the steam from seeping in?" The students believe this is true, so I ask them how we should do it. They decide to use several bottles and balloons and securely tape each new balloon onto its bottle before placing it in the pot of hot water. We try four different setups with bottles of different sizes and shapes and balloons of different sizes and shapes. In each case, the balloon still inflates.

Though it contradicts their own ideas, the children enjoy watching the balloons inflate. Each time I repeat the experience, I invite the students (especially those who most firmly believe that the source of the air is *outside* the bottle) to stay close and observe. Jamila looks hard for steam seeping into the bottle. Mike holds the balloon tightly around the neck of the bottle to make sure it is airtight.

More tries . . .

After the four additional tries, I ask the students what they are thinking now. At this point, they have mostly decided that the air inflating the balloon must be air that was already inside the bottle. We make connections to hot air balloons as well as where we find the hottest room in our houses in the summer. They are coming, in their own way, to the science idea that I was hoping they would grasp.

. . . and a new idea emerges

In these experiments, as the students have noticed, the balloon always fills gently and partially with air; it does not inflate completely. I ask them why they think the balloon did not blow up more than it did. Some students remark that it could fill up only with the air that was already in the bottle when we started.

Finally, to help them build on their understanding, I ask myself what other demonstration I can do to show that there is air in the bottle. We proceed to take the balloon off the bottle and invert the "empty" bottle in a basin of water. The water does not rise in the bottle. Some students say that it must be the air in the bottle that keeps the water from entering.

Another experiment

HELPING STUDENTS CONSTRUCT MEANING

Reflecting on the story   Let's reflect on what happened in Ms. Hudson's class when I helped the students experiment with bottles and balloons. As you'll recall from Chap-

LEARNING SCIENCE IS SOMETHING STUDENTS DO, NOT SOMETHING THAT IS DONE TO THEM.

—National Science Education Standards

An active process of making meaning

ter 2, reflection requires careful thought; it is a deliberate process of thinking about an experience and achieving your own understandings of the concepts and principles embedded in that experience. By reflecting on the bottles-and-balloons story, we can clarify some of the ideas introduced earlier in this book.

As we noted in Chapter 1, the view that learning is an active process of knowledge construction is part of a family of theories called *constructivism.* Basic to constructivism is the necessity for individuals to put together thoughts, interpretations, and explanations that are their own personal constructs (von Glaserfeld, 1992; Yager, 1991). This means that in order to understand a concept, individuals *must* be engaged in the active process of making sense of their experiences. This is an essential principle of constructivism: the belief that simply "knowing" something does not constitute understanding. Instead, we gradually come to understand ideas as we turn them over in our minds and reflect on our experiences.

As you have already gathered, this theory has important implications for methods of science teaching. Understanding how the learner constructs meaning helps us to create environments that stimulate our science learners.

Meaning Making Versus Language Recall

Science educators have often relied on merely *exposing* students to basic science concepts and skills. What we know is that the student can repeat the language we request, but we are not sure that there is any deep meaning in this sort of science knowledge. *Grasping terminology is not the same as understanding the concept.* Many studies indicate that just because students can memorize concepts well does not mean they can apply those concepts in a real-life context (Yager, 1991; Brook & Brooks, 1993).

Rote answers . . . understanding

When I asked the students in Ms. Hudson's class what was in the bottle, all the students apparently agreed on the answer: "air." Nevertheless, I could tell that many were not comfortable identifying the air in the bottle. For some, it was just a rote answer, something they had been told was true but hadn't fully incorporated into their own thinking. Although air is all around us and seeps into everything, it is a difficult concept for young students to understand. They can blow air onto their hands, feel the force of a breeze—but, still, it is a tricky concept.

Further, in the bottle-and-balloon experiment, the method by which heat energy travels in liquids and gases is called convection. Ms. Hudson or I could have introduced this term and asked the students to memorize

Language and Meaning: One Child's Scientific Experience

Richard Feynman, a Nobel Prize winner in physics, told the following story about an incident in his childhood that demonstrated the difference between knowing terminology and knowing concepts.

We used to go up to the Catskill Mountains for vacations. . . . On the weekends, my father would take me for walks in the woods. He often took me for walks, and we learned all about nature, and so on, in the process. . . .

[Other boys in the area wanted to do the same.] So all fathers took all sons out for walks in the woods one Sunday afternoon. The next day, Monday, we were playing in the fields and this boy said to me, "See that bird standing on the wheat there? What's the name of it?" I said, "I haven't got the slightest idea." He said, "It's a brown-throated thrush. Your father doesn't teach you much about science."

I smiled to myself, because my father had already taught me that that doesn't tell me anything about the bird. He taught me "See that bird? It's a brown-throated thrush, but in Germany it's called a halzenflugel, and in Chinese they call it a chung ling and even if you know all those names for it, you still know nothing about the bird. You only know something about people; what they call that bird."

[His father continued:] "Now that thrush sings, and teaches its young to fly, and flies so many miles away during the summer across the country, and nobody knows how it finds its way," and so forth. There is a difference between the name of the thing and what goes on.

The result of this is that I cannot remember anybody's name, and when people discuss physics with me they often are exasperated when they say "the Fitz-Cronin effect"—and I ask "What is the effect?" and I can't remember the name.

Reprinted with permission from Richard P. Feynman (1968), "What Is Science?" *The Physics Teacher,* 7 (6): 313–320. Copyright 1968 American Association of Physics Teachers.

the definition. But even after learning the term, they might well have been unable to relate it to the balloon and the bottle.

Keep in mind this distinction between acquiring language and truly creating meaning. There is nothing wrong, of course, with knowing the term **convection,** but knowing it as a term and understanding it in real life are different. As Piaget (1964) noted, the construction of ideas is an *active process,* both physically and mentally.

Prior Knowledge

The philosopher John Dewey said that concepts should be viewed as "known points of reference by which to get our bearings when we are plunged into the strange unknown" (1933, p. 153). That is, concepts don't just sit there in our minds; they help us interpret and deal with new situations. In turn, these new experiences help us refine our concepts. I like to think of our concepts as old friends—ideas we come to know as we grow and that we refine and revisit as we add new understandings to our repertoire. Our repertoire of such understandings is our **prior knowledge.**

Did the students in Ms. Hudson's class have prior knowledge? To a casual observer, it may not seem that they knew very much about the matter at hand. Nevertheless, the prior knowledge they possessed was extremely important. They knew that boiling water gives off steam, for example, and they certainly understood some of the characteristics of steam.

As I conducted my experiments, I was really interested to learn that several students thought the balloon inflated because of the steam from the just-boiled water. If I had merely told them something different, they might have nodded at me but failed to derive any real meaning from what I said. Instead, because my interest in their thinking was genuine, the students knew that I valued their ideas and respected their prior knowledge. This leads us to our next crucial point.

> Prior knowledge—our repertoire of familiar concepts

Valuing the Students' Thinking

Valuing the students' own ideas is a way of communicating to them that they are important members of the class. That is important in itself, but there is more to it. When their ideas are genuinely valued, students begin to see themselves as *knowers.* Only as knowers can they construct new meanings by building on their prior beliefs and ideas.

This is a vital principle to grasp. Creating an atmosphere in which there a sense of trust makes it possible to help the students reflect on their new experiences and use those reflections to modify their prior knowledge. Without such trust, the students are not likely to do much serious thinking about the science experiences in which you engage them. They may memorize a "fact" or two, but their underlying conceptions may remain unaffected.

When you inquire about their thinking, you need to be ready for many types of responses. *All* of the responses have value since they provide in-

> Valuing students' ideas → students' reflection → new ideas

Valuing students' thinking helps them to reflect upon their ideas. Here, the teacher listens as the students explain their thoughts.

Carol Palmer

Avoiding the "teacher game"

sight into the students' thinking. If you tell the children your thinking first, they will never feel comfortable sharing their own ideas. The students will immediately think the way you think, or at least they will say they do. This is called playing the "teacher game," and even young children are good at that.

The most important role you will play when you teach science is to find out what the students think and help them build on their knowledge. If they have difficulty reaching a "correct" scientific understanding, that is okay. Listen to the children's ideas and reflect on them. Ask the students questions that help you understand the nature of their prior knowledge. Encourage them to:

ALL OF US LEARN THROUGH OUR EXPERIMENTS. DO WE GIVE OURSELVES THE FREEDOM TO EXPERIMENT, THE FREEDOM TO BE WRONG, THE FREEDOM TO BE RIGHT . . . FOR A LITTLE WHILE . . . UNTIL WE TURN OUT TO BE WRONG AGAIN?

—*Jacqueline Grennon Brooks and Martin Brooks*

- Write about and draw what they have observed.
- Create stories and poems about their science activities.
- Plan other similar experiments to test their personal ideas.

Such experiences will encourage them to come up with their own ideas and have those ideas reflected in the discourse of the classroom.

Science activities and experiments need to provide students with the freedom to say exactly what they think—even if that means it is the freedom to be wrong. In Part Two of this book, you will find many examples of this teaching strategy.

The Rewards of Valuing a Student's Thinking

One preservice teacher wrote the following passage in her science journal, describing her encounters with Jonathan, a five-year-old at a preschool.

The children were outside playing. A few children came up to me and said, "Melanie, look, there is grass growing on the ground."

I said, "Wow, I wonder why the grass is growing."

They replied, "Because it is getting warm out."

I said, "Why is it getting warm out?"

A little boy named Jonathan answered, "Because it is getting to be spring."

"Well," I said, "can you tell me what happens in the spring?"

"It gets warm," the children explained. "Flowers grow. We can go outside and play."

Then they ran off to play, but the boy named Jonathan stayed by me. We walked around and saw some flowers growing. He told me that at the tip of the stem was a bud. I said I wondered what was in the bud. So we took the tip off the stem and Jonathan took it apart. As he did so, he put all the pieces in my hand. He was amazed, just as I was, to see what really was inside.

As we discovered all the pieces in this flower, he was speculating about the function each piece had in the development of the flower. Many of the other children surrounded us to see what we were doing. After we finished with the bud, I asked Jonathan what he thought we should do with it. He said to sprinkle the pieces on the ground "because then maybe more flowers will grow." So we sprinkled them on the ground. [Notice that Melanie did not immediately tell Jonathan he was wrong.] After that, Jonathan was so excited that he wanted to see if any other buds were growing. As we walked around the playground, we searched for something else to observe. We found a tree with flowers. We took apart the bud of this tree and came up with some observations, such as what color we thought the flowers would be, how many flowers would grow from this one bud, and why this bud was different from the other one we had just looked at.

It was interesting to see how a five-year-old became so wrapped up in the observation. . . .

[*Four days later:*] All of a sudden Jonathan came running up to me with much excitement. He gripped my hand and told me that he had to show me something. He then directed me to some more buds he had found. As we took one apart, like we had done before, I realized what an impact I must have made on him. For him to feel free enough to observe on his own and know that he could come to me to talk about it meant the world to me.

Have you had any similar experiences with young children? If so, you may want to write about them in your own science journal.

Reprinted by permission.

Mediating the Students' Learning

What does it mean to "mediate"?

If our interest is helping students construct meaning, then we have to teach in ways that facilitate that process. The challenge is to help students delve deeply into their thoughts and expand their own thinking about an idea. Our role therefore becomes that of a **mediator.** In everyday terms, a mediator serves as a go-between of some sort, often in helping people resolve their differences by bridging the gaps between different points of view. A teacher who is a mediator helps students bridge the gap between their initial understanding and the deeper knowledge they can build as a result of the lesson. The teacher can do this in a variety of ways, but the process usually begins by exposing students to new experiences and helping them probe their own thinking.

In the bottle-and-balloon lesson, I was helping the students probe their thinking by inviting them to try the experiment again, in different ways. I facilitated that process by providing additional materials and becoming the mediator of their ideas. It wasn't easy. The students who believed that

One-Way Versus Two-Way Communication

The role of teacher as mediator is very different from the traditional concept of the teacher as simply a *provider* of knowledge. One way to understand this difference is to think of it in terms of communication.

In traditional science teaching, the process has been accomplished as *one-way* communication: the teacher talks, the students listen. But if we really believe that our students are knowers—that they bring their prior knowledge of the world to our classes—then we can never think of science teaching as a one-way process. Teaching becomes a *two-way* conversation, a dialogue in which the teacher explores a student's thinking and responds, and the student then explores the teacher's thinking and responds. In fact, when groups of students become involved, the conversation often involves three or four people or more.

With this style of communication, the teacher helps students flesh out their ideas and make meaning from them. In the end, this type of science teaching yields a deeper, more important, and more valid understanding of science ideas.

IN MY ENTIRE LIFE AS A
STUDENT, I REMEMBER
ONLY TWICE BEING
GIVEN THE
OPPORTUNITY TO
COME UP WITH MY
OWN IDEAS, A FACT I
CONSIDER TYPICAL
AND TERRIBLE.

—*Eleanor Duckworth*

Students' theories

Alternative
conceptions—stages in
a journey

Dealing with
alternative
conceptions

the air inflating the balloon came from outside the bottle had a hard time realizing that anything was already in the bottle when we started. Although they said "air" when asked what was in the bottle, they were merely being correct, not truly understanding the concept.

Hence, I encouraged them to test their own theories. By "theories" in this context, I mean their proposed explanations for a natural event. In science, we generally use the term *theory* to mean a belief about a science idea that has a lot of observable evidence to support it. By students' theories, we refer to their beliefs based on their own understandings at a particular level in their cognitive development. These may not be full-fledged theories in the scientific sense, but they can lead to genuine theory building as students construct further meaning.

As the story demonstrates, different children construct meaning in different ways, and sometimes their ideas amount to **alternative conceptions**—ideas that are not scientifically accurate but represent a step toward full understanding of a concept. As you will see in the next chapter, alternative conceptions are like train stations—places the children pass through on their journey to fuller understanding. Obviously teachers need a thorough grasp of science concepts on a given topic in order to know when the children's own ideas represent steps in the path toward fuller understanding. As a teacher, you achieve that grasp by exploring science experiences and science concepts one topic at a time—and by locating your inner scientist. (Do not worry; it is doable and very rewarding.)

When students reveal alternative conceptions, teachers need to evaluate those ideas and take action. This "action" is what your role as mediator is all about. Valuing the students' ideas does not mean you must let them go unchallenged. Instead, you invite your students to examine their conceptions from multiple perspectives—making connections between the experience and something else in their memory, or trying the experiment again in a different way. Thus I asked Mike if he had ever seen steam go through a bottle before. I asked the students, too, if they thought making the seal tighter would prevent the steam from seeping into the balloon, and I guided them to explore that possibility. Through such means, we help them refine their ideas so that their thinking matches their own enlarged experience. In the next chapter, we will see further examples of the teacher as mediator.

Now let's look at another example of students' constructing meaning. The following story shows second graders investigating the process of evaporation. I first heard this story while attending a workshop entitled Benchmarks for Scientific Literacy (AAAS, 1993). A different version of it appears in the *National Science Education Standards* (National Research Council, 1996).

S C I E N C E   S T O R Y

Two Dishes of Water

A second-grade teacher set out two dishes of water. They were identical in every way except that one dish was covered and one was not. After the teacher and students established all the ways in which they were the same, the teacher left the two dishes out for the weekend. When the students returned to class Monday morning, they noticed that the water level in the uncovered dish was lower than the water level in the covered dish. The teacher asked the students to think about what had happened to the water.

Some water vanishes

"Maybe the hamster got out and drank the water," one student suggested. "Can't he knock the top off his cage?"

Did the hamster do it?

"Maybe the custodian was thirsty this weekend," another student remarked.

"How can we test these ideas?" asked the teacher. The students secured the hamster cage lid, spoke to the custodian about the dish of water and the experiment, and tried leaving the two dishes out again.

When, again, the uncovered dish had lost some water, the class reached a conclusion. They decided that if you leave two dishes of water out and one is uncovered, then some of the water in the uncovered dish will seem to disappear!

FROM EXPERIENCES TO MEANING MAKING

Dewey (1933) talked about concrete experiences as the ground and generator of the thinking process. Remarking that experience would lead to the understanding of ideas and the capacity to conceptualize, he argued for hands-on learning fifty years before it became a catch phrase. This move-ment from objects to ideas is what rooting science learning in concrete ex-perience is all about. Some researchers have used the term *situated cognition* or *situated learning* to refer to the notion that the situation or context of learning is critical and that real learning takes place only when situated in real-world activities (see, for example, Gage & Berliner, 1998). Overall, the theoretical foundation for rooting science teaching in mean-ingful, concrete activities is strong.

On the other hand, the story of the two dishes of water reminds us that experience doesn't automatically produce what we consider accurate sci-entific ideas. The notion that the water in an uncovered dish "seems to dis-appear" is only a small step beyond the conception that "the hamster did it." The challenge for the science teacher is to help the students move from this stage to the concept of *evaporation*. How does such movement occur?

Well, we could *tell* the students that the water didn't really disappear—that it *changes state* into a gas that we cannot see. I think about the second graders I have known who memorized the idea that "evaporation is the process by which a liquid changes to a gas." That phrase could be written down and tested—and still the student might not relate it to the water in the two dishes. The difficulty that these particular second-grade students had in conceptualizing that water vapor, a gas, is really there is not unlike the difficulty the third graders had in believing that there really was air in the bottle. (See Figure 3.1 for a related example.)

So what do we do? We can begin by asking ourselves, "What have they noticed?" It appears that these second graders know that *something* has happened to the water, something that they cannot see. Therefore, one way to lead them to a deeper understanding may be to make the process of evaporation more visible. Here are some possible steps in that direction:

Moving toward understanding

1. If you take a large bell jar, place 5 centimeters of water inside, and then close it, you create a humid condition in the jar. Students will be able to see moisture in the form of water vapor condensing on the inside of the jar. This is one way to begin talking about the idea that the water did not just disappear—it only looked that way.

Building on what students have noticed

2. After they identify the humid condition in the jar, help the students make connections between moisture in the jar and moisture in the air. The students should probably have already formed some understanding of solids, liquids, and gases as part of their prior knowledge.

Introducing terminology at an appropriate time

3. You could introduce the term *evaporation* by asking, "Would you like to know what scientists call it when water changes from a liquid to a gas?"

FIGURE 3.1

· ·

Young students' ideas about the states of matter. These excerpts are from the science journals of two students in the same class. It seems that student A had difficulty grasping the concept of a gas. Student B gave a correct example of a gas, but may still not have fully understood that gases have weight and take up space even when they are invisible.

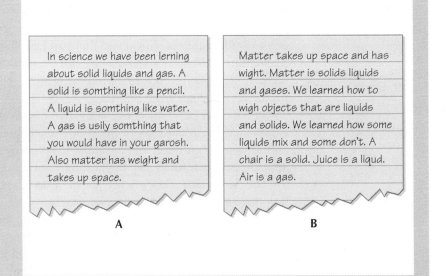

A

In science we have been lerning about solid liquids and gas. A solid is somthing like a pencil. A liquid is somthing like water. A gas is usily somthing that you would have in your garosh. Also matter has weight and takes up space.

B

Matter takes up space and has wight. Matter is solids liquids and gases. We learned how to wigh objects that are liquids and solids. We learned how some liquids mix and some don't. A chair is a solid. Juice is a liqud. Air is a gas.

Then you could set out more water-filled containers of different sizes and encourage the students to observe what happens.

4. You could invite the second graders to think about ways that evaporation occurs every day. For example, when wet clothes dry on a clothesline or when we wave wet hands through the air to dry them off, we are watching water droplets seemingly disappear; actually, the droplets are evaporating, changing to a gas. Referring to such everyday experiences brings classroom meanings into another dimension of the students' lives.

Connecting with everyday experience

Through steps like these, we give the students opportunities to think about phenomena and come up with their own ideas. Then we encourage them to sort out their thinking, revisit their ideas, and experiment with them.

· ·

CHILDREN AS KNOWERS ·

This chapter has demonstrated that scientific knowledge requires an active process of meaning construction, involving explorations of our own

Being invested in
seeking answers

questions and testing of our own ideas. For children, this means that they become invested in seeking their own answers. The answers they seek enable them to fit their new ideas into their already existing conceptual framework—their prior knowledge.

I am reminded of a story by my friend Judy Logan (1997). At age five, Judy's niece, Maren, said to her, "Teachers know everything." Judy answered, "No, teachers know *some* things. Nobody knows everything."

As Maren clung stubbornly to her belief that teachers know everything, Judy explained, "But there are things you know that your teacher doesn't know."

"Like what?" Maren demanded.

"Like what it is like to be Molly's big sister," Judy told her, "and my five-year-old niece."

It's an amusing story, but Maren's initial opinion of teachers reminds us that schools are places where, often, children do not feel like "knowers." They feel that teachers know everything and they, as children, have little to contribute.

A shift in thinking

Constructivism as a theory of learning provokes us to find teaching strategies that honor students' ideas and create safe, open spaces for students to create their own meanings. This is not the traditional notion of science teaching. For many people, it requires a shift in thinking. Rather than simply passing down or transmitting information with one-way communication, the teacher creates a forum for two-way or multiway exchanges, mediating the students' progress on the path to their own deeper understanding. The teacher's role as mediator includes the following:

Useful ways to
mediate learning

- Engaging students in concrete experiences
- Encouraging them to express their ideas about what they observe
- Listening seriously to those ideas, considering how they are based on the students' prior knowledge
- Encouraging the students themselves to reflect on their ideas
- Offering additional possibilities, such as other possible avenues of thought for them to explore, further questions to consider, and new connections with their daily experience
- Suggesting new experiments and providing opportunities for students to carry them out
- Repeating the process as the students have further experiences, so that they continually build new ideas and refine their old ones

The following poem in two voices helps to express this shift in thinking about the teacher's role. The voice on the left side represents some traditional notions about science teaching. The right side represents the kind of approach that I hope this book will foster:

I teach science to children.	I do science with children.
I provide information.	I provide experiences.
I seek to control the lesson.	I watch the lesson unfold.
I like it when my students explore *my* thinking.	I like it when my students explore *their* thinking.
My students ask what they should be observing.	My students trust their own observations.
My students memorize facts from each lesson.	My students reflect on ideas from each lesson.
My students know the science ideas for the next test.	My students connect the science ideas to their lives.
My students know that science has all the answers.	My students know that science helps find some answers.
I teach science to children.	I do science with children.

Constructivist teaching is like doing science

In many ways, using the theory of constructivism to influence your science teaching practice is a lot like doing science in a research lab. You pose your problem, you observe and "listen" to the experiment, and you revise your process as you respond to the changes in the experimental system. You reach your conclusions gradually, and sometimes you repeat the experiment. Hence, as you proceed to learn about how to teach science, you further develop your scientific self.

Key Terms

convection *(p. 58)*
prior knowledge *(p. 59)*
mediator *(p. 62)*
alternative conception *(p. 63)*

Resources for Further Exploration

Electronic Resources

Building an Understanding of Constructivism (1995). *Classroom Compass,* 1(3), **http://www.sedl.org/scimath/compass.** This introduction to the third issue of the online journal *Classroom Compass* offers a fine overview of constructivism. Begin at the web site listed and follow the links to Volume 1, Number 3.

Crowther, David T. (1997). The constructivist zone. *The Electronic Journal of Science Education.* **http://unr.edu/homepage/jcannon/ejse/ejse.html.** Follow the links to Volume 2, Number 2, of this electronic journal. Crowther's editorial offers a historical perspective on constructivism and an overview of its impact on current practice. The issue as a whole is devoted to the "science wars"—the controversy about methods of teaching science.

Print Resources

Duckworth, E. (1991). Twenty-four, forty-two, and I love you: Keeping it complex. *Harvard Educational Review*, 61(1): 1–24.

Duckworth, E., Easley, J., Hawkins, D., & Henriques, A. (1990). *Science Education: A Minds-on Approach for the Elementary Years.* Hillsdale, NJ: Lawrence Erlbaum Associates.

Gallas, K. (1995). *Talking Their Way into Science.* New York: Teachers College Press.

Hawkins, D. (1965). Messing About in Science. *Science and Children*, 2(5).

Kaner, E. (1989). *Balloon Science.* Reading, MA: Addison-Wesley.

Logan, J. (1997). *Teaching Stories.* New York: Kodansha International.

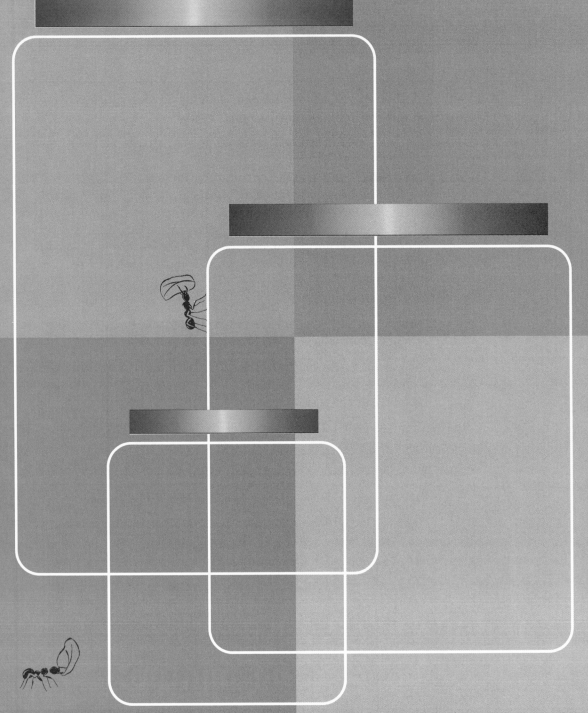

Doing Science with Children: Inquiry in Practice

I n Part Two, we explore many types of science activities that help children come up with their own ideas and seek answers to their own questions. We investigate these activities by means of *science stories*, narrative accounts of actual experiences that teachers, including myself, have had while doing science with elementary school children.

In Chapter 4, we look in more detail at the teacher's role as mediator of the elementary science experience. Next, we explore a classroom "science circus" that demonstrates what I mean by *inquiry* in elementary school science. In later chapters, we delve into other aspects of science teaching, such as making connections with students' daily lives, sustaining inquiry over time, model making, and using the typical "science box" or kit that is often supplied with the curricular materials.

In addition to questions about how to teach, all of the chapters in Part Two examine particular topics in science. Three broad classifications of science content—life science, physical science, and earth science—are frequently used to guide the development of science instruction at the elementary school level:

- In *life science,* students and teachers explore characteristics of living things and their interactions with the nonliving environment.
- In *physical science,* students and teachers look at properties of objects and materials, forms of energy, and motion of objects.
- In *earth science,* students and teachers investigate properties of earth materials, objects in the sky, and changes in the earth and sky.

To find material on a particular topic, see the List of Science Topics on page XX.

The activities and commentary in Part Two are just a sampling of what occurs in elementary school science. The activities are designed to be repeated, revised, and modified to fit the needs of your own students. The commentary should remind you to enjoy what you are doing, believe in it, and become a science learner, *separate from and together with* the children.

Look especially for the *Expanding Meanings* sections following the stories. The subsections called *The Teaching Ideas Behind This Story* and *The Science Ideas Behind This Story* will help you think about the concepts that are illustrated by, and embedded in, the stories. Also, *Questions for Further Exploration* can take you to the next level of thinking about experiences like the ones described in the stories. And *Resources for Further Exploration* offer both print and electronic material to expand your knowledge base in each area of science teaching.

4 The Teacher as Mediator

FOCUSING QUESTIONS

- When you think of a coach, what images come to mind?

- Have you ever been on a sightseeing tour bus, listening to the guide?

- Do you prefer to be left alone to explore or to be guided in your explorations?

In Chapter 3 we talked about the teacher's role as mediator. In this chapter we explore that role further by examining a science experience in a third-grade class. When we provide meaningful science experiences, we are always asking ourselves how much free exploration we should offer the students and how much we should guide and coach them. In the following story Mr. Wilson has to choose appropriate times to mediate the students' experience by involving himself in their explorations, making suggestions, and pointing out ideas.

SCIENCE STORY

Icicles

It is an icy-cold winter morning in the Northeast. It snowed two days ago, and the temperature has plummeted to well below the freezing point of water. Ice and snow cover everything.

On his way to school in this urban community, Mr. Wilson notices icicles hanging from the edges of roof lines. The icicles glisten in the sun. They are of varying lengths and thicknesses. He reaches up and breaks off some extra-long ones and brings them to his third-grade class.

The students arrive bundled up with scarves, gloves, and hats in addition to their heavy winter coats. They settle into their seats, and after the morning business, Mr. Wilson reaches to the ledge outside the classroom window, where he has been storing the icicles.

He shows them to the children. "Where do you think I found these?" he asks.

The children call out all the places that they have seen icicles this morning. Some of them noticed icicles hanging from tree branches; others saw icicles on roofs and awnings. Still others didn't seem to notice any icicles this morning. As a group, the children are excited that Mr. Wilson has brought some icicles to class. He describes how he reached up to a low roof and gently pulled them off, trying not to break them.

"What are all the things you notice about these icicles?" Mr. Wilson asks.

The children respond in various ways: "They're long." "They're cold." "They're cloudy." "They're hard." "They're pointy." "They will start dripping when they melt." "You can hurt someone if you stick it into them!"

"Let's explore these things before they melt," says Mr. Wilson. "What are your questions? What do you want to find out about these icicles?"

Some children are interested in knowing how long it will take for them to melt. Others suggest different questions:

"Will they weigh the same when they melt as they do now?"

"When they melt, will they be the same color?"

"Can we make them back into an icicle after we melt them?"

"Do they taste good?"

Mr. Wilson notices . . .

WITH THE YOUNGEST CHILD, THE OBJECTS OF THE WORLD . . . ARE PREGNANT WITH SUGGESTIONS, AS THEY GIVE RISE TO QUESTIONS.

—Maxine Greene

The students notice . . .

The children have questions.

"Doing Scientific Stuff" Versus "Doing Science"

Recently, as part of a professional development class, an English teacher spent some time observing a science lesson in the classroom of a teacher named Philip. Many of the outward signs of good science learning were there: the students were doing hands-on experiments in small groups and dealing with some sophisticated concepts. But as she observed the class, the English teacher noticed that something crucial was missing. Here are some of the eloquent comments she wrote:

Yes, students were busy in groups of their choosing. But Philip came around to each group, suggesting directions for them to take in working on the experiments he had designed. Yes, students were doing scientific experiments, making highly conceptual physics come alive without needing to be able to read and write and speak English at a level commensurate with the abstraction of the physics they were learning by doing experiments. But some students did not contribute much to the group efforts. . . . And Philip coached and coaxed the preparation for the presentations I saw as well as the actual presentations themselves. He orchestrated, played first violin, and conducted the orchestra. He wanted them to do more by themselves, but, according to an interview I conducted, they weren't getting it this cycle. So he directed them into participation.

All along I had the uneasy feeling that this wasn't science. It was doing scientific stuff, but it wasn't doing science. My metaphor for doing science is my experience in a biology lab when I had finished all the teacher's textbook/workbook experiments and had time left over. I decided to conduct my own experiments with my dead, splayed frog. I did heart massage with a warm, 20 percent saline solution poured into its chest cavity. I got the heart to beat rhythmically on its own. I had posed my own question which was of interest to me, had designed my own experiment, and had conducted it with no sure idea of what would happen, just to see if what I had read and heard about saline solution were true.

What Philip's students were doing was playing out a deterministic game. His experiments were a different form of the workbook or worksheets common to my discipline area. They say to students, "Learning is self-contained. It has a definite beginning, middle, and end. We can know the end before we start, or at least we can control the outcome. It can be easy if you only follow the correct steps and do it my way. I know best. I know what you are supposed to learn. Follow my directions."

Though the writer says in the first paragraph that Philip "coached" the students, this does not seem to be the kind of coaching done by Mr. Wilson in the icicles story. Think about the key differences. Do you agree that Philip's students were "doing scientific stuff" but not really "doing science"? For them to be doing science—in the sense that this observer means it—what would have to change?

Reprinted by permission.

Mr. Wilson's thinking: Notice that Mr. Wilson has asked the students to think about what they want to know about the icicles. He knows that what is really important is the students' questions—the ones they cannot answer yet, ones that compel them to search for answers.

Why are the students' questions important?

Mr. Wilson engages the children in a discussion of their questions. They talk about why it may not be a good idea to taste the icicles (some children think they may be dirty). Most of the children want to make the icicles melt. Some children want to place the icicles on the classroom heater. Some want to light a candle (Mr. Wilson would do that for them) and hold the icicle over it. Some children want to place the icicles on the window ledge, where the sun is streaming in.

At this point Mr. Wilson returns the icicles to the outside ledge temporarily. He asks the children to divide up into their usual science groups. Each group will get an icicle to work with, he says, and each group should decide in advance what to do with it before it melts. As a suggestion, Mr. Wilson repeats one of the ideas the students themselves proposed: "Let's do find out," he says, "if our icicles weigh the same when they melt."

Why does Mr. Wilson let the children decide?

As the children are deciding on their procedures, Mr. Wilson visits each group and coaches them. He points things out to them and asks them to explain what they have come up with. He prods them, leading them with his questions, coaxing and coaching their planning.

"What should we do to find out if the icicles weigh the same after they melt as they do now?" he asks various groups.

"Weigh them, before and after," the children figure out.

Mr. Wilson's thinking: Although this is an impromptu lesson, Mr. Wilson has an objective in mind. He's hoping the students will begin to learn what melting really is. The question about an icicle's weight before and after melting was an important one, so he stimulates their attention to it. He also guides them in framing their investigations so that they will be able to explore the changes that occur when the icicle melts. He points things out and inquires about their plans. He is somewhat like a tour guide or a coach. Unlike a tour guide, however, he does not explain the details of all the sights. Instead, he listens to the students' impressions and asks for whatever meanings they may construct.

Being a coach

It turns out that one group cannot decide what to do. Mr. Wilson asks them what they have thought about so far. One student explains that "Jamie wants to make something that will keep the icicle from melting," while the rest of them (three other children) want to time how long it will take to melt.

Mr. Wilson thinks about this dilemma and asks Jamie what question she has in mind. Her question is, "How long can I keep the icicle cold?" She

Jamie's different idea

wants to wrap the icicle up with materials from the classroom and put it in a lined shoebox. "Why are you interested in doing that?" asks Mr. Wilson. Jamie replies, "I want to see if I can keep it cold long enough to take it home."

"Okay," Mr. Wilson says, "why doesn't Jamie do that with one icicle? The rest of you can take another icicle and work on your plan to melt it."

Mr. Wilson's thinking: Although his main focus in this lesson is on melting, Mr. Wilson doesn't force all the children to work on that. He knows that Jamie is becoming invested in exploring her own question, so he allows this to happen, being flexible in his ideas about group work.

When the groups have decided on their questions and procedures, Mr. Wilson takes out the scales—the double-pan balances that are common for measuring mass in elementary grades. He distributes an icicle to each group, with one extra for Jamie.

But before the students begin to weigh the icicles, he asks them, "How will you collect the water while your icicle is melting?" (Some of the icicles are already dripping.) The children have to consider the problem. He invites them over to the science supply table and asks for suggestions. They look at the plastic cups, aluminum foil pie pans, shoeboxes, and assorted plastic containers. Someone suggests that they allow the icicle to melt in an aluminum pie pan. Mr. Wilson encourages this idea since the pie pans are flat and can accommodate the icicles without tipping over.

> Encouraging a good solution

Now he asks the children to weigh their icicles. As their standard masses, the children use teddy bear counters: thick plastic figures, about 5 centimeters long, shaped liked teddy bears. (Standard unit masses may be in gram units or in any other convenient units—for example, paper clips or pennies.)

In their small groups, the children take turns with the double-pan balance. Some children weigh the icicle and the pan together. Others weigh just the icicle. Mr. Wilson observes each group and does not give specific instructions. The children realize that the icicles have to be weighed quickly, before they melt.

> No specific instructions

Mr. Wilson's thinking: Notice that Mr. Wilson does not direct the students in how to weigh their icicles. He wants to allow the children the freedom to experiment—the freedom to be wrong and then to explore further until they are right. He watches, he suggests, he leads a bit, and then he lets go—until (as we will see) he needs to intervene. He is mediating their experience.

After the weighing, the children proceed with their different plans for melting their icicles. One group, which has weighed the icicle in the pan, places the pan on the sunny window sill and checks it every few minutes

Mass and Weight

In elementary school science, students learn some basic ideas about matter. Matter can be defined as anything that has weight or mass and takes up space (as opposed to energy, for example, which has neither of those qualities).

Mass refers directly to how much matter is in an object. Often the term *weight* is used interchangeably with *mass*, but in scientific usage, the two are different. *Weight* means the gravitational pull that the earth has on an object.

Since the gravitational pull of the earth is relatively constant, mass and weight are essentially the same on earth. But if you weigh an object on earth and then take it to the moon, it will weigh only about one-sixth as much because the moon's gravitational pull is about one-sixth as strong as the earth's. The object's mass, on the other hand, will be the same in both places.

Even on earth, the nearer a body is to the center of the earth, the greater the downward pull of gravity on the body and the more the body will weigh. The farther a body is from the center of the earth, the less the body will weigh.

until the icicle melts. Then the group members weigh the melted icicle in the pan and compare the two measurements they have made:

icicle + pan = 20 teddy bears
melted icicle + pan = 20 teddy bears

Other groups follow different methods of melting and come up with similar results in the weight test.

Predictably, however, the groups that have weighed an icicle by itself, outside the pie pan, run into trouble. For instance, one of these groups proceeds to melt the icicle in the pan on the heater and then weighs the resulting water in the pan. This group concludes that the melted icicle weighs more than the solid icicle. Now Mr. Wilson intervenes. "What about the pie pan that you have the water in?" he asks. "Doesn't it weigh something?"

The children look at him.

He explains, "You weighed your icicle on the balance scale all by itself, but you weighed the water in the pan. If you weighed the icicle all by itself, then you would have to weigh the water all by itself."

Thinking about this, the children come up with a new plan. They ask Mr. Wilson if they can try the experiment again with another icicle. "Yes,"

A "right" answer

A "wrong" answer

I THINK AND THINK FOR MONTHS AND YEARS. NINETY-NINE TIMES, THE CONCLUSION IS FALSE. THE HUNDREDTH TIME I AM RIGHT.

—*Albert Einstein*

he says, and gives them a new one to work with. He has been keeping extra icicles outside the classroom window.

A new attempt

*Mr. Wilson's thinking: Mr. Wilson intervenes when he realizes that the children
have come up with an* alternative conception, *an idea that has been constructed
from their experience but is not scientifically accurate. He sets them on a path that
will help them build a more accurate idea.*

   *It is interesting, though, that Mr. Wilson does not tell them to find the mass of
the pie pan by itself and subtract that from the total mass of the pie pan and water. The children have not thought of that option, and Mr. Wilson decides to allow
them to redo their experiment rather than impose his procedure on them. This is
an important decision: finding the right balance between guiding the procedure
and allowing students to experiment for themselves. Here we see another facet of
the teacher as mediator.*

Finding a balance

The group that is redoing the experiment begins by weighing the icicle
in the pan this time. They decide to let the new icicle melt in the pan on a
table in the back of the room. This icicle takes longer to melt, so they weigh
the resulting water after lunch. This is their result:

icicle + pan = 15 teddy bears
melted icicle + pan = 15 teddy bears

   Once all the groups are finished, Mr. Wilson again engages the entire
class in discussion. The students all agree now that you have to weigh the
icicle in the pan at the beginning, and then you have to weigh the melted
icicle in the pan at the end. They also agree that an icicle's weight does not
change when it melts.

Agreement at last—
and surprise

   Although they agree with this conclusion, most of them are very surprised by it.

   "Well, what did you think would happen?" Mr. Wilson asks.

   The children say that the melted icicle looked so little compared to the
original icicle that they thought it would weigh less (see Figure 4.1).

   "What other melting experiments can we do?" Mr. Wilson wonders
aloud. The children suggest using ice cubes to see if the same thing will
happen. Mr. Wilson promises to bring ice cubes to class the next day.

And what next?

   Meanwhile, Jamie keeps her icicle wrapped tightly in its shoebox and
takes it home that afternoon.

FIGURE 4.1

....

Like the students in Mr. Wilson's class, young children frequently think that water derived from melting ice will weigh less than the original ice because the water *looks* smaller.

E X P A N D I N G   M E A N I N G S

....

The Teaching Ideas Behind This Story

■ The way Mr. Wilson used the icy weather to engage the students in a science activity reflects an important connection between the outside environment and the activity within the classroom. Making this connection is a very important part of doing science with children. Nature is all around us, and frequently it presents itself through changes in the weather. These activities may be thought of as *informal science learning experiences* as contrasted with the more formal, prescribed science curricula that we will explore later in this book.

■ Clearly Mr. Wilson was engaging the children in a constructivist learning activity. Notice the way he listened to their ideas about icicles, weighing, and melting and then encouraged them to try out their notions. This method helps students recognize what they already know and build on it. For example, the students knew that the icicles would melt if kept indoors. They had their own ideas about places where the melting might happen quickly—the heater in the classroom and the sunny window ledge. Their prior knowledge also helped them choose an appropriate container to place each icicle in before they melted it. Because Mr. Wilson acknowledged them as knowers, they could proceed to experiment with confidence.

This teacher is interacting with a small group of students. When teachers act as mediators, they question, guide, and give suggestions.

Richard Hutchings/Photo Edit

■ Nevertheless, Mr. Wilson did not ignore what they were doing or let them experiment entirely without help. Instead, he *mediated* their experience by giving them suggestions and guiding their activities.

■ There are those who will think that Mr. Wilson should have told the children at the beginning that they needed to weigh the icicles in some kind of receptacle. This kind of instruction may be time efficient, but it does not promote the development of the children's own ideas. He wanted the children to explore on their own to learn the best way to measure their icicles. For that reason, he created an atmosphere in which the children had the opportunity and *intellectual freedom* to investigate their notions. He also gave them the tacit message that it is okay to be "wrong."

■ When one team of students decided that the water weighed more than the original icicle, Mr. Wilson knew that this was an alternative conception that could become a stopping-off place en route to deeper meaning. It was, in effect, raw material for him to work with in helping the children construct deeper knowledge.

■ Mr. Wilson did not know in advance exactly where the icicle experience would lead. He did not impose a rigid procedure. He was intending that the students learn something about melting. But Jamie, the student who kept the icicle cold, would be learning something about insulation. He did make sure that each group explored the question of weight before and after melting—a question that the students themselves had raised in the beginning when he prompted them for ideas.

The Science Ideas Behind This Story

■ The activity in this story addresses what happens when matter *changes state,* in this case from a solid to a liquid. The students observe that when a certain amount of matter changes state, its mass does not change. Changes in state are associated with different amounts of energy, but the amount of mass remains constant before and after the change of state. Each group of children added heat energy to its icicle so that it would melt. The icicle absorbed the heat energy. That caused its particles to move faster and spread farther apart, turning the solid into a liquid. But the additional energy did not affect the mass.

■ This activity helped students understand that when weighing the same object in different forms, the method used is very important. There is a key general principle here. In a science experiment we always ask, "What are we keeping the same? What are we changing?" Scientists call these **constants** and **variables,** respectively. The students were changing the state of

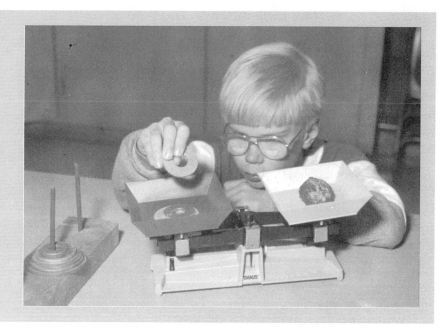

This student is weighing a solid using washers as uniform masses. How many washers weigh the same as the solid object on the double-pan balance? In scientific activity, we are often seeking answers through measurement.

Tony Freeman/Photo Edit

matter; therefore, to make this a fair test of the effect on weight, the conditions under which the two states were weighed had to be the same. A **fair test** in science requires that we keep all the experimental conditions the same (constant) except the one we are testing for (the variable).

Questions for Further Exploration

- In what ways do you think the icicle activity stimulated the students' thinking process?

- What prior experiences do you think the children were relying on to construct new meanings?

- What science ideas did the teacher need to know before engaging children in this activity?

- What materials did the teacher rely on to design this activity? What other materials might have been useful?

- What connections from this lesson could you make to literature or social studies?

- How might you integrate technology into this activity?

Resources for Further Exploration

Electronic Resources

Constructivist Classrooms. **http://129.7.160.115/INST5931/constructivist. html.** A summary and discussion of a book by J. G. Brooks and M. G. Brooks, *In Search of Understanding: The Case for Constructivist Classrooms* (Alexandria, VA: Association for Supervision and Curriculum Development, 1993). Includes a list of pointers on constructivist strategies.

Just Think: Problem Solving Through Inquiry. Videotape series available from the New York State Education Department, Office of Educational Television and Public Broadcasting, Cultural Education Center, Room 10A75, Albany, NY 12230. Offers an excellent view of teachers as mediators in real classrooms.

A Private Universe Video. A 20-minute cassette about basic science concepts and alternative conceptions. Available from the Annenberg/CPB Math and Science Collection, Corporation for Public Broadcasting, 901 E Street NW, Washington, D.C. 20004. Call 1-800-965-7373 to order.

The Weather Unit. **http://faldo.atmos.uiuc.edu/WEATHER/weather. html.** Lesson plans relating to weather, with projects in math, art, reading, and other fields, as well as science. See also the weather sites listed in the Expanding Meanings section of Chapter 5.

Print Resources

Dyasi, H. (1995). Is there room for children's ideas in the elementary school curriculum? In William Ayers (ed.), *To Become a Teacher: Make a Difference in Children's Lives.* New York: Teachers College Press.

Hoover, E., & Mercier, S. (1990). "Melt an Ice Cube," an activity described in *Primarily Physics* (1990). Fresno, CA: AIMS (Activities Integrating Mathematics and Science) Education Foundation.

Museum of Science. (1992). *The Weather Kit.* Boston. This is a multidisciplinary curriculum kit from the Museum of Science in Boston. Very student centered and interactive, it has great ideas for weather experiments.

Naturescope. (1989). *Wild About Weather.* Washington, DC: National Wildlife Federation. Provides good background information about the causes of weather.

Pyramid Film & Video (1988). *A Private Universe: An Insightful Lesson on How We Learn.* Santa Monica, CA.

MEDIATION AND ALTERNATIVE CONCEPTIONS

Predetermined, predictable lessons

Many science teachers, unlike Mr. Wilson, offer students a completely predetermined type of science lesson, precisely planned and predictable. Such lessons are often designed to control the outcomes; the teacher knows beforehand exactly where the students should end up. There are obvious attractions to this method. After all, wouldn't you like to know exactly what your students are supposed to learn and how to get them there?

But the problem with imposing strict procedures on children when they are doing science is that the teacher does not learn what the children really think or how they construct meaning. Without learning what they think, it is extremely difficult to help them change or develop their thinking. Many years of research have revealed that most people have significant misconceptions about nature—that is, they believe that there are reasons for natural events that are actually not so (see, for instance, Pyramid Film & Video, 1988). It appears that even many years of schooling do not help.

The staying power of inaccurate ideas

STANDARDIZED TESTS OFTEN DO NOT PICK UP STUDENTS' MISCONCEPTIONS.

—*Robert Yager*

Alternative conceptions—a better term

Changing alternative conceptions

Everyone has misunderstandings

People cling to their misconceptions. Even university science and engineering majors have many misconceptions about science.

It is difficult to know exactly why this is so. The exact pathway for learning any concept, including a misconception, remains obscure. What science education researchers do know is that standardized tests often do not pick up students' misconceptions. Often students will score well on standardized tests and yet be unable to change the experience-based interpretations of nature that they acquired prior to instruction (Yager, 1991). Remember the children in Chapter 3 who thought that steam caused the balloon on the bottle to inflate. Even though they said "air" was in the bottle, they interpreted the event as if the bottle were empty.

Misconceptions are better thought of as "alternative conceptions" because it is important to foster a tolerance of these ideas in order to provide students with the experience necessary to change them. Let's consider the icicles story again. The children discovered that the mass of the solid icicle and pan was the same as the mass of the melted icicle and pan. Most of the children, seeing the small puddle of water left when the icicle melted, thought it must weigh less than the original icicle. This is because, unlike other liquids, water expands when it freezes, taking up more space, and conversely it shrinks when it melts.

The idea that the water would weigh less than the icicle was contradicted by their experience. Without this experience, they might have memorized the fact that water expands when it freezes without really knowing in practice what that meant. It is only by becoming aware of children's embedded ideas about the natural world that we can begin to change their alternative conceptions. By providing opportunities for them to express their own ideas, talking about those ideas, and guiding them through tests of the ideas, we can mediate the development of more sophisticated and more accurate concepts. This role will take on deeper meaning for you once you experience it in your own classroom.

For young children, some common alternative conceptions involve the idea of the earth as a cosmic body. Research on this subject was done as early as 1976 by Nussbaum and Novak, who learned that second graders who had received six audio-tutorial lessons on the earth in space had several notions, including: (1) the earth is flat; (2) the earth is round but limited in space by a floor and a ceiling; and (3) the earth is round, bounded by limitless space, but on it things fall "down." All three of these notions, the researchers found, persisted until around age ten or beyond.

The alternative conceptions that learners bring to the science learning experience cut across age, ability, gender, and cultural boundaries. They persist regardless of who the students are and where they live (Driver,

Some Strategies to Help Change Persistent Alternative Conceptions

- Listen to the students' ideas.
- Honor their thinking.
- Ask students to explain their thinking: "That's interesting—what makes you think that?"
- Alter experiments. For example, do the same experiment again with different materials.
- Ask students to devise a plan to demonstrate that their alternative conception works: "Could we plan an experiment to see if that works?"

1989). An award-winning videotape, *A Private Universe,* created by Project STAR (Science Teaching Through Its Astronomical Roots), shows that even Harvard graduates hold fundamental misunderstandings about the solar system. This is an example of just how difficult it is to change or remove misconceptions developed in childhood. It argues for the importance of teaching strategies designed to reveal these alternative conceptions and provide repeated experiences that help change them.

Now let's see how Mr. Wilson followed through on his lesson with the icicles.

SCIENCE STORY

From Icicles to Ice Cubes: The Next Day

The following morning, the children in Mr. Wilson's class come to school ready to see if a pan of ice cubes will weigh the same before and after melting. This experiment was the children's own idea, and it generates curiosity and profound interest.

Before addressing the ice cubes activity, however, Mr. Wilson gives Jamie a chance to present her experiment to the class. She was the student who was determined to explore how long she could keep her icicle before it melted. She had wrapped it in aluminum foil and paper towels before

Jamie explains her results.

placing it in a plastic bag and putting it in a shoebox to take home. She explained to the class that her icicle did last, but it became a lot smaller. When she got home, she put it in the freezer.

Now Mr. Wilson brings out three trays of ice cubes from the school refrigerator. Weighing the tray of ice cubes before and after melting will be a repeated experience, another visit to the concept of masses remaining the same when state of matter changes. This is an example of assessing how well the children can *apply a concept* they learned the day before. Once again, Mr. Wilson is engaging them in solving a problem, setting up an experiment, testing their own ideas, and drawing some conclusions.

Can they apply the concept?

But the students now are also interested in Jamie's type of experiment. They want to know if they can try different techniques to preserve the ice cubes. "Sure," replies Mr. Wilson. One ice cube tray can be used for melting, and the others can supply the cubes for the new experiments. Mr. Wilson sees that the class's further explorations can lead to ideas about the concept of insulation and the losing and gaining of heat energy. In this way, one student's question has become an entire class's experiment.

The class wants to explore further.

FOLLOWING THE STUDENTS' QUESTIONING

One good question breeds another

Do you see how the approach to learning in Mr. Wilson's class parallels the scientific process itself? In professional science, one experiment often leads to another experiment because somebody in the research team has a new, but related question. In the same way, Mr. Wilson's students, this time stimulated by Jamie, follow their own questioning from one exploration to another.

In the rest of the chapters in Part Two, watch for the ways in which teachers listen to the students' thinking and work with their ideas, mediating the children's experience. Keep in mind that when we do science with children, both the children and the teacher are science learners.

Key Terms

mass *(p. 77)*
constant *(p. 81)*
variable *(p. 81)*
fair test *(p. 82)*

5 The Science Circus: Using the Skills of Scientific Study

- Do you ever predict the day's weather from your early morning observations?

- Can you remember diagnosing a sick pet or a child or a sibling on the basis of the symptoms you noticed?

- Did you ever vary just one ingredient in a recipe and come up with a new dish?

- Do you ever shop at the supermarket by arranging your grocery list according to the aisles in the store?

People who regularly engage in scientific activity use certain skills that help them gain information about nature and natural phenomena. As we noted in Chapter 1, these skills, such as observing and classifying, are also important in our daily lives, though we often do not "code" them as scientific. Employing these skills on a planned and regular basis is what makes scientific activity different from ordinary activity. It is the essence of what we mean by the scientific process of **inquiry,** and the skills themselves are generally known as **inquiry skills** or **process skills.**

In this chapter, we explore a multistation classroom environment that provides children and teachers with science activities that require the use of certain process skills in order to resolve a question or solve a problem. By a "station" I mean simply a desk, a table, or any other special area in which an investigation is described and materials are provided. By examining this environment and thinking about how it works, you can become more familiar with the ways in which children develop process skills and use them to solve problems.

What Is Inquiry?

The following passage from the National Science Education Standards *offers an important explanation of what scientists and educators mean when they speak of the process of "inquiry."*

Scientific inquiry refers to the diverse ways in which scientists study the natural world and propose explanations based on the evidence derived from their work. Inquiry also refers to the activities of students in which they develop knowledge and understanding of scientific ideas, as well as an understanding of how scientists study the natural world.

Inquiry is a multifaceted activity that involves making observations; posing questions; examining books and other sources of information to see what is already known; planning investigations; reviewing what is already known in light of experimental evidence; using tools to gather, analyze, and interpret data; proposing answers, explanations, and predictions; and communicating the results. Inquiry requires identification of assumptions, use of critical and logical thinking, and consideration of alternative explanations.

From National Research Council, *The National Science Education Standards.* Washington, DC: National Academy Press, 1996, p. 23.

THE SCIENCE CIRCUS

If you have ever been to a circus, you have noticed how many different things are going on at the same time. In three-ring circuses, entertainment acts of different varieties occupy the rings in the center, while jugglers, clowns, and people on stilts walk around the periphery. It is quite a busy place. One hardly knows what to look at first. For children, these events create a world of excitement that they want to explore.

A colleague of mine introduced me to a multistation classroom activity that she labeled a "science circus" because many activities were occurring simultaneously. I said, "When I think of a circus, I think of disorder." "On the contrary," my colleague remarked, "it takes a great deal of organization to plan an activity where students are all at different places at the same time" (Abder, 1990).

Lots of activities—but not disorder

A **science circus** is not quite as complicated as a real circus. Contained in a room, the science circus consists of several stations at which the visitors are asked to perform certain tasks and record their results or reactions. It is circus-like because many different activities are going on simultaneously, but it is far from chaotic.

Providing personal context

Each time I set up a science circus, it looks different. I try to use materials common in daily life and create a scenario that invites students to explore them and come up with their own ideas. Often, too, I situate the activities in a personal context related to my own experience—a context I hope the students will recognize and share. It is always important to provide a rich context for the science stations so that you are not pacing the students through science activities just for their own sake. The experiences must be constructed in a way that makes a connection to a larger idea or a personal story.

Arousing a need to know

In the following story, I describe one typical science circus that I helped create, but keep in mind that the potential for different activities is practically infinite. Notice, too, that the stations try to arouse in the students a *need to know*, giving them a strong reason to use their science skills. As you explore these stations, you may be reminded of instances in which you have satisfied your own need to know by using the skills of science.

After seeing how the science circus works, we will explore the particular skills involved in each of the activities. As you read, think about how *you* would describe the skills that the students are developing.

SCIENCE STORY

The Circus Comes to Mount Holly

I am visiting Ms. Markon, a new teacher at Mount Holly Elementary School. I have been her teaching supervisor, and today we are setting up a science circus in her fourth-grade class. We have used the students' lunch break to prepare the tables in the room. The students are aware that they will be returning from lunch to find that their room has been transformed.

When the students come in, they discover several tables, each with a different object or collection of objects and an index card with questions on it. They are delighted with the transformed classroom and eager to begin. Because it is important for students to share ideas with each other, Ms. Markon asks each to work with a buddy as they explore the stations in the science circus. Each pair of students is assigned to a different station to

A classroom transformed

Two children explore the hollow in a log from an old tree trunk. What could have caused that hole to form?

Alma Rocha/Hofstra University Day Care

start. They are instructed to visit each of the stations, spending between five and ten minutes at each stop. As they explore the stations, I have the opportunity to interact with the students.

An Old Log

At one station, the table holds an old log from my backyard. It comes from a birch tree that was felled in a storm several years ago. The log is very dried, but its thin, whitish bark is still intact. It is about 2 feet (60 cm) long and has a diameter of about 5 inches (12.5 cm). One end is rough where it broke off from the rest of the trunk. The other end is smooth because the log was sawed for firewood. In the middle of one side of the log is a large, oval opening, probably made by the gradual pecking of a large-beaked bird.

Ms. Markon and I do not tell the students everything we know or surmise about the log. Instead, the index card on the table next to the log reads as follows:

> This is a piece of a tree trunk that I found in my backyard. Can you explore this piece of trunk and try to figure out what happened to it?
>
> What are all the things you observe about this tree trunk? What are some of your ideas about how it got to look this way?

The children handle the small log and record their ideas. Working in pairs, they talk about the log and how they think it got to be here in our science circus. One girl takes a marble from a neighboring shelf and drops it into the hole in the log. When I notice her doing that, I walk over to watch more closely.

"Oh!" she exclaims. "I'm sorry I took the marble!"

Surprised that she believes she has broken some rule, I reassure her that I think she has a great idea and that I'm just curious about what she wants to find out.

"I was just trying to see if the hole in the log went all the way through," she says. "It doesn't."

My thinking: The girl's reaction was a dramatic illustration of how rule-bound students become after only a few years in school. I reminded myself to rejoice in those moments when students use their creativity to explore.

The students readily notice the differences between the two ends of the log. "Somebody sawed this," says one. "This looks like it broke off," another student suggests, looking at the other end of the log. "I bet a woodpecker did this," still another student offers, looking at the large hole where the marble has been rolling around. They draw pictures of the log in their science journal. They also record the title of the station and their observations and ideas about the log.

The Soda Can

On another table we have placed a soda can that was left in a car on a very hot summer day. The heated gas bubbles in the soda exploded the lid of

Just playing around— or being creative?

THE FASCINATION OF ANY SEARCH AFTER TRUTH LIES NOT IN THE ATTAINMENT BUT IN THE PURSUIT.

—*Florence Bascom, first woman to earn a doctorate in geology in the United States*

the can at its seams and puffed out the bottom. The can is empty now, but it looks quite unusual. The lid is still attached at one point on the edge. The index card at that table reads:

On a very hot summer day, I left my unopened soda can in the back of my car. Several hours later, when I returned to my car, the soda was all over the car seat and this is what the can looked like. What do you think happened?

Listening to students brainstorm about this can is very interesting. Once," says one child, "my mom left a soda can in the freezer by mistake, and this is what it looked like." Another child remarks, "It looks like it exploded." The students record their various ideas in their science journals.

Brainstorming

My thinking: I am interested to learn whether the students make a connection between temperature and the behavior of liquids in a can. I'm using a real story from my experience, one that I hope they can relate to. Asking new questions about daily-life experiences is my way of engaging them in the task of problem solving.

Crazy Rocks

In a large plastic bowl on one table, Ms. Markon has arranged her rock collection, consisting of several different kinds of rocks. She has collected these rocks from many different places, and they are of various colors, textures, sizes, and shapes. The index card on the table reads:

Effect of Temperature on Liquids

At the soda can station, students have the opportunity to learn what happens when we heat or freeze a liquid. When a carbonated liquid is kept in a very hot place, the carbon dioxide gas dissolved in the soda water becomes less soluble as the temperature increases. Heat energy increases the speed at which the liquid and gas particles move. The faster the motion of the gas particles, the more often they hit the sides of the container and the greater the pressure. This pressure can eventually burst the seams of the can.

As one of the students in the story realized, the same kind of thing can happen when a soda can is frozen. Even though liquids contract when they are cooled, water-based liquids like soda water start forming crystals when they begin to freeze, and the ice crystals occupy more volume than the original water. As the carbonated liquid occupies more space, it can burst through the seams of the can. This usually does not occur when soda cans are left in the freezer or out in the icy cold weather because there is usually a little space in the soda can that the freezing soda expands into. Many containers of water based liquid, when filled to the top and sealed do burst when frozen. Hence, if the students inferred that the can was left either in a freezer or in a very hot place, they were correct.

When astronauts brought rocks back from the moon, the first thing the scientists did was to group them according to similar characteristics. This is MY rock collection. You may have a rock collection too. I started collecting rocks when I was a small child.

When you explore these rocks, what do you notice? In what ways are they the same? In what ways are they different? How many categories could you place the rocks into? Where do you think I found these rocks?

The children love rummaging through these rocks. It is a fairly large collection, and they come up with categories like "bumpy," "smooth," "light colored," "dark colored," "shiny crystals," "no crystals," "stripes," and "no stripes." Some of these rocks may be ones the students have never seen before.

Making up categories

My thinking: Although many students could not relate to some of the rocks in the collection, most could relate to the idea of having a collection—at least an informal group of rocks that they had picked up here and there because they liked them.

The Leaky Faucet

This station is at the sink in the classroom, where water is dripping slowly from the faucet. (If your room has no sink, you can use a large jug of bottled water with its own spigot and a basin to collect the dripping water.) Nearby are a stopwatch, beakers, and graduated cylinders (see Figure 5.1). The card reads as follows:

You may have seen a leaky faucet in your own home. Measure and write down the amount of water that drips from the tap in one minute. Work out how much water will drip in one hour. How would you find out how much water would drip in one whole day?

As I watch the pairs of students, I remind them to write their plan in their science journals before carrying it out. Then I continue to watch and listen as they solve the problem.

My thinking: I am interested in understanding how the students will figure this out. What will they do? What units will they use? What will their decisions tell me about their prior experience and their ways of thinking?

FIGURE 5.1

A dripping sink or water jug makes a simple but effective station in a science circus, stimulating students to observe, measure, and calculate.

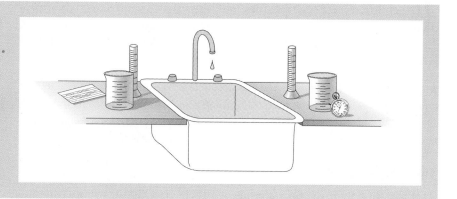

Many of the students are surprised to learn that as many as 7,200 milliliters of water can drip from a leaky faucet in one day if only 5 milliliters drip out in one minute. The next day, Ms. Markon brings in empty liter bottles, and the students help her stack seven of them up to see what 7,000 milliliters looks like. Several of them suddenly realize why their parents nag them to turn the water off.

Surprise—and understanding

The Penny in the Pie Pan

At this station there are some pencils and a few pennies on the table, plus an aluminum foil pie tin half filled with water. The students read the following:

> Many of us like to swim. Have you ever noticed that your hands and feet look different when they are under water?
>
> Today we are going to submerge a penny in the pie tin of water. Before doing that, take your pencil and trace the penny in your science journal. Now draw what you think the penny will look like when you put it in the water. Now place the penny in the water. What do you notice?

The children trace their pennies with their pencils and sketch their predictions. When they see the pennies in the water, they are often surprised.
"I didn't know it would look so big," one boy exclaims. They record their
observations in their science journals.

More surprise

*My thinking: At this station Ms. Markon and I want to engage the children in
an activity that invites them to look at ordinary materials in a different way. Pennies, water, and pie pans are commonplace items. Making the connection to swimming is another way to provide a personal context for the activity.*

Three Types of Soil

At this station there are no materials at all. This is a "thought experiment."
The card reads as follows:

Imagine you have three different types of soil. How
would you determine which soil holds the most
water? What will you keep the same? What will you
change?

The students spend a long time at this station. They brainstorm a way
(or several ways, sometimes) to figure out the problem. When they decide
on a way to figure it out, they write it in their science journals. Here is an
example from two of the fourth graders:

**Use three cups. Put three holes on the bottom of each cup and stand each
cup in a pan—like a pie pan from the last station. Add each soil to a cup
[one type of soil per cup] and then put the same amount of water in each**

cup. Wait 1 hour. Measure water that drips out the bottom. The cup that doesn't drip out too much water has the best soil.

I ask this pair of students, "How big is each cup?" I want them to understand that we need to keep the size of the cups the same. Then I ask, "How much soil do you put in each cup?" I want them to see that exactly the same amount of each soil should be used. They decide to use half a cup of each soil. Next I ask them, "How will you measure the water that drips out?" The two children explain that they will pour it from the pie pan into a measuring cup.

On my next visit, I bring in three types of soil, paper cups, and measuring cups, and these two students try their experiment. It works! We examine the soils for size of particles and color. We determine how the soils are the same and different.

Another pair comes up with a different solution. They will also use paper cups, they decide, but they won't punch holes in the bottom. Instead, they will add the same amount of water to each type of soil and wait fifteen minutes. Then they will take three sheets of paper toweling, and on each sheet they will empty one of their cups, dumping out both the soil and the water. They have reasoned that whichever paper towel is the wettest will contain the soil that can hold the least amount of water.

"How much water will you add to each cup?" I ask.

They haven't thought about that—just that each cup would get the same amount of water. They think some more about their plans. On my next visit, they carry out their experiment. They have decided to add one-half cup of water. But they find that their cups are too small for that amount, so they need to add less. They try again.

My thinking: I particularly like thought experiments, especially ones involving experimentation with materials that the students can readily imagine. In this case I used soils because most children have had plenty of experience with dirt! Also, of course, samples of three types of soil were easy for me to bring to the classroom on my next trip.

The Weather Station

This station has a computer set up to access an Internet weather site (**http://www.weather.com**). This and other weather sites on the World Wide Web can provide statistics, photos, and explanations of unusual weather events. At this station the card reads:

Margin notes:

Guiding with questions

A different plan

On-line investigations

Accessing the weather map from the Internet (http://www.weather.com) **allows students to collect weather data from distant places in an instant.**

Courtesy, The Weather Channel

Compare the air temperature and weather conditions for San Francisco, California; Chicago, Illinois; and New York City. In what ways are they the same? In what ways are they different?

Which of these cities would you prefer to be in today? What city on the national weather map has your favorite weather today?

The children use the Internet site to gather data that they record in their science journals. They need to manipulate the computer to obtain the information they are seeking.

My thinking: The relatively simple task of recording data from a web site is one way in which elementary school students can become familiar with using computers as a tool for scientific study. This same activity engages the children in map reading, giving them experience with another significant tool. I like to personalize the activity by asking the students to choose a city that has their favorite weather.

E X P A N D I N G   M E A N I N G S

The Teaching Ideas Behind This Story

■ The science circus stations provide opportunities for students to practice using scientific process skills. While doing this, they also gain practice with investigations that generate some original thinking. Scientific inquiry is not just about doing; it is also about critically thinking about what you do.

■ Notice how curious the girl was at the station with the small log. When she acted on her curiosity by placing a marble in the hollow, she was momentarily unsure if she had done the right thing. Often youngsters are timid about trying out their ideas, fearing they will be "wrong." Teaching science depends on overcoming that fear.

■ At the station with the thought experiment about three types of soil, there were at least several possible "right" answers. Challenging students to solve a "How would you find out?" problem invites multiple responses as students personalize the problem and critically explore potential solutions. Notice, through my questions, how I prodded them into setting up a fair test, by pointing out how important it is to keep the experimental conditions the same. Notice, too, that at a later time I brought different soils to class. Thought experiments can be frustrating to the students if the experiments cannot actually be tried.

■ The science circus engages students in using scientific process skills without necessarily naming the skills. In fact, this chapter itself has not yet named all the skills involved. (It will shortly.) But when, you may ask, should you name the skills for your students? Generally, naming and subsequent discussion of the science process skills are appropriate by the end of third grade and the beginning of fourth grade. At that point, like the process skills themselves, the names should be used regularly by both teacher and students so that their meanings become situated in the students' experiences.

The Science Ideas Behind This Story

■ The activities of the science circus are designed to develop particular science process skills. Though most of the stations involve more than one skill, the principal skills used at each station are highlighted below.

■ At the station with the log, we are asking the students to use the skills of observation and inference. An **observation** is all the information we can gain about an object or an event by using our senses. An **inference** is a reasonable statement about an object or an event that is based on an observation. We make observations with our senses and use our minds to make inferences—to really think about something. For instance, what happened to this tree trunk? Why does it look this way?

■ An inference may or may not be provable. In science, if an inference or a guess is stated in a way that allows it to be tested, it is called a **hypothesis.** Typically scientists plan ways of testing their hypotheses to find out if they are true. In this way, a hypothesis may eventually become a **theory**—an idea that has been tested and (to some significant degree) proved. But even though they have been "proved," theories are often in a state of revision as they are tested and retested and further evidence accumulates.

■ The same process skills apply to the soda can station: observation and inference.

■ When the students explored Ms. Markon's rock collection, we were asking them to observe in a particular way. This process skill is **classifying**—sorting objects or ideas into groups according to similar properties. Classifying is a basic way that human beings try to make meaning of the natural world and all its diversity. We use classifications in all our thinking. For instance, by identifying the object in front of you as a book, you are classifying it—connecting it with a group of similar objects.

■ The leaky faucet station involved **measuring,** another process skill. Here the students had the opportunity to work with tools that help extend our abilities to observe. At this station, they used a stopwatch and graduated cylinders and beakers, helpful for measuring time and volume. Other useful instruments for measurement are balance scales, spring scales, thermometers, and meter sticks.

- When they wrote down their measurements, the students were **recording data.** In elementary school science, data include many different types of information, and students can record them in words, numbers, graphs, or drawings.

- At the penny and pie pan station, the children were asked to **predict** what they thought the penny would look like under water. Then, after they dropped it in the pie pan, they **compared and contrasted** their observations with their predictions.

- At the station without materials, which we called a thought experiment, the students were asked to **plan an investigation.** In order to do this, they had to recognize that testing ideas involves *controlling variables.* That is, they had to ask themselves, "What do I keep the same, and what do I change?" As Chapter 4 pointed out, when we do science, we have to design experiments in which all variables but one are controlled. So if we want to find out which soil holds the most water, we need to keep the volume of the water and the amounts of the soil and all other experimental conditions the same, changing only the type of soil.

- At the weather station, students were again recording data after accessing the information on the Internet. In addition, using computers as a tool to collect information is such an important skill that it may deserve to be called a "process" skill in its own right. This skill has become vital for scientists in many theoretical fields, especially when direct observation is not possible because the objects to be observed are too small or too far away. In these cases, computers make models of the objects of study, and data are generated as scientists manipulate the models.

- Educators differ in the exact way they list and define the process skills. But the ones highlighted in this section are included in virtually every list. These process skills are summarized in Table 5.1.

Questions for Further Exploration

- Think about a science circus station that you would like to create—perhaps something very different from the stations described in this chapter. If possible, plan it for a particular class with which you have had

TABLE 5.1   **The Process Skills**

Observation	Gaining information about an object or an event by using all your senses
Inference	Making statements about an observation that provide a reasonable explanation
Classifying	Sorting objects or ideas into groups based on similar properties
Measuring	Determining distance, volume, mass, or time by using instruments that measure these properties (such as centimeter sticks, graduated cylinders, scales, stop watches)
Recording data	Writing down (in words, pictures, graphs, or numbers) the results of observations of an object or an event
Predicting	Guessing what the outcome of an event will be on the basis of observations and, usually, prior knowledge of similar events
Comparing and contrasting	Discovering similarities and differences between objects or events
Planning an investigation	Determining a reasonable procedure that could be followed to test an idea (listing the materials needed, writing out the procedure to be followed, and identifying which variables will be kept the same and which will be changed)

some experience. Then ask yourself the following questions about your plans.

■ What process skills will be necessary for students to complete the task?

■ How engaged will the students be in their own thinking? Will your station provide opportunities for them to generate their own questions? How can you promote their thinking?

■ When they see the materials at your station, what do you think your students will be most interested in learning about? (Sometimes I have an idea about what I think is important to explore in science, and the children have a different idea.)

■ How can the station be relevant to your students' lives?

- What sorts of things do you expect to learn about the children from their questions? And how will this knowledge help you plan further science activities for them?

Resources for Further Exploration

Electronic Resources

Excite: My Channel. **http://my.excite.com/weather**. At this weather site, students can look up local weather reports as well as explore the weather in other parts of the country.

Science Process Skills Benchmarks. **http://www.math.clemson.edu/faculty/ dmoss/aophub/Benchmarks/Process_Skills.html**. This web site gives a detailed description of the types of student behaviors that can be expected when using the science process skills in grades K–3 and also in grades 4 and 5.

SEGway: Science Education Gate-way. **http://www.cea.berkeley.edu/ Education/SII/SEGway.** This web site has a variety of tools and lesson plans in earth and space science. The weather section offers a real-time weather map, links to other weather pages, plus additional resources for content learning about meteorology.

Weather.com. **http://www.weather.com.** Another weather site, with easy access to maps.

Print Resources

Bosak, Susan V. (1991). *Science Is . . .* Ontario, Canada: Scholastic Canada and the Communication Project. A terrific sourcebook for elementary school teachers, this volume includes over 450 activities that can lead to exciting avenues of exploration with students.

GEMS. (1997). *Sifting Through Science.* Berkeley, CA: Lawrence Hall of Science. This curriculum guide, designed specifically for early childhood education, explains in detail how to set up learning stations at the primary grades.

Ostlund, K. (1992). *Science Process Skills: Assessing Hands-On Student Performance.* Menlo Park, CA: Addison-Wesley. Although its activities are designed to be assessments of student process skills, the book also has useful suggestions for a science circus.

SCIENCE IN THE CLASSROOM AND IN EVERYDAY LIFE ....

Process skills you use every day

The skills of scientific inquiry are also the skills that are used to live thoughtfully throughout life. If you answered "yes" to any of the Focusing Questions at the beginning of this chapter, then you have been using process skills to make decisions that affect you every day. For instance, packing a suitcase for a trip requires lots of classifying: casual wear, dress wear, nightwear, toiletries, undergarments, shoes. Decision making about products, the weather, our health, and the state of our nutrition requires careful observing, inferring, comparing, and contrasting.

When these skills are exercised in a particular way, our inferences can become hypotheses. Remember Rose in Chapter 2? When Susan had the interesting idea about why the balloon didn't pop, Rose said, "How can we test this possibility?" Susan's idea was framed as a hypothesis and subjected to a test. Testing ideas in such a planned and consistent fashion is the foundation of scientific inquiry.

Building process skills over time

Through activities like the ones in our science circus, young children begin to build their process skills, the foundations for their understanding of the natural world. They learn to observe closely, sort things into categories, ask how things are the same and how they are different. Even very young children can make predictions—what will happen if? They come up with tentative explanations for their observations, and these inferences change as they get older. Children begin measuring by using string, centimeter sticks, and balance scales. They weigh objects with uniform masses like the teddy bear counters in the icicles story of Chapter 4. As they mature, they begin to use standard units of measurement for distance, mass, and volume. By third grade, many children can plan investigations, recognize variables, and control the experimental conditions. At all levels, students are recording data, first through drawings and later through words, graphs, and numbers.

Engaging students' minds

Remember that the science circus experience, like all other science activities, is a success only if it generates children's questions and promotes further thinking. Their minds as well as their hands have to be applied. They will not learn process skills from mere manipulation of objects. The real learning occurs as a result of their brainstorming, discussion, and reflection. In the next chapter we'll look more deeply into ways of promoting children's learning by connecting science experiences with their own lives and environments.

Key Terms

inquiry *(p. 87)*
process (inquiry) skills *(p. 87)*
science circus *(p. 88)*
observation *(p. 100)*
inference *(p. 100)*
hypothesis *(p. 100)*
theory *(p. 100)*
classifying *(p. 100)*
measuring *(p. 100)*
recording data *(p. 101)*
predicting *(p. 101)*
comparing and contrasting *(p. 101)*
planning an investigation *(p. 101)*

6

Making Connections: Science in the Students' Own Environment

FOCUSING QUESTIONS

■ How can informal science learning experiences shape your knowledge of science?

■ What natural event has been making headlines recently?

■ How does the area in which you live adapt to changing climate conditions?

■ What items do you tend to collect? Do they include anything from nature?

Many elementary schools follow a **formal science curriculum** that has been developed at the state or local level or that has been commercially developed and purchased by the school district. We will examine some typical science curricula in Part Three of this book. These formal curricula have generally been put together with the guidance of the *National Science Education Standards* or the AAAS *Benchmarks* (see Chapter 1).

Unfortunately, teachers often use only these formal guidelines to design students' science experiences. They therefore limit the nature of the science experiences to those addressed in formal curricula. Although those experiences may be good—and may make some topical connections to students' lives—they do not include experiences that are *unique* to a particular class in a specific geographic locale.

Science experiences that are unique and very personal make up what I call the **informal science curriculum.** In this chapter, you will see that this informal curriculum can be extremely important. You can think of it as a way to integrate spontaneous natural events from the school's locale and science experiences that the students bring from home. In Chapter 4, you saw the informal science curriculum at work when Mr. Wilson brought icicles into class to teach about changes in matter. Students can also study local objects from nature and even make collections of the plants found outside the school.

This chapter explores a variety of informal science learning experiences, especially the ways they can be developed through a classroom science corner and through field trips. First let's look at some of the reasons that it is so important to make these connections between science and your students' daily lives.

DIVERSITY WITHIN AND WITHOUT

Look outside your classroom window.

Wherever your school is located, you will find enormous diversity in nature. Often you can see some of it just by looking out your classroom window. You may find yourself amid the rich deciduous forests of the suburban or rural Northeast, or close by the vast cornfields of the Midwest, or near the rocky shores of a coastal community. In the Southwest or Southeast, you may be surrounded by the dramatic flora and fauna of arid or tropical regions. Even at an inner-city elementary school, you may find trees in planting squares in the sidewalks or at least a dandelion breaking through the cracks.

How do different students respond to the surroundings?

You probably have a great deal of diversity among your students. The students will notice different things in nature and respond to them in different ways. Consider the little dandelion in the sidewalk, for example. One student may see it as a sign of the stubborn persistence of plant life; a student with relatives in the suburbs may see it as a weed that ought to be exterminated. One student may come from a family that serves dandelion greens at a meal; others probably have no idea that people would eat such a thing. One may think the dandelion flower is beautiful; another may think it ugly or not even recognize it as a flower. Whatever their particular reactions, your students' unique ways of experiencing nature will provide a rich tapestry of science experiences that can become part of the life of your classroom.

By taking advantage of this diversity inside and outside your class—seizing the opportunities to make interesting connections between students' lives and the natural world around them—you can make your classroom a dynamic, vital place. Moreover, building these links stimulates learning in two major ways:

1. It helps students see the larger picture.

2. It counteracts the traditional alienation of many people from science.

Seeing the Larger Picture: Content Plus Context

All students learn better if they can relate their school experiences to their daily lives. Recall what Chapter 1 said about the need to fit new concepts into our prior knowledge—a concept that many of the science stories have already demonstrated. Particularly in science education, however, many adults can recall learning the content without any context to which they could relate it. There was no larger picture, so to speak. As a result, these students "learned" science by memorizing decontextualized facts for the test and then forgetting them afterward (reported in science autobiographies; see Koch, 1990).

Problems of content without context

Today, even formal science curricula generally strive to find ways that students, especially young children, can relate science ideas to their lives. And the best curricula acknowledge that students' lives encompass a wide variety of activities, cultures, and home environments. By linking science with their own, individual "larger pictures," we not only help youngsters learn better, but also help them use science ideas to understand events in their daily lives more fully. As we discussed at the beginning of the book, we are living in an increasingly complex world—one that requires its citizens to be observant, critical, and thoughtful about their immediate environment. Those who understand scientific ideas in the context of their daily lives have a head start in dealing with the issues that their lives present.

Linking science with a larger picture

From Alienation to Inclusion

Various researchers have investigated the reasons that science has often alienated students, especially females and members of minority groups. A major culprit, studies have found, is the absence of personal connections between science and students' everyday lives. For example, minorities and females often have a negative perception of the usefulness of science in "real life," and this attitude contributes to their lack of participation in science activities (Clewell, Anderson, & Thorpe, 1992; Koch, 1993a). Conversely, by relating science to these students' daily lives, we can create school science experiences that will invite their participation.

Lack of connection → lack of participation

Helping students to see scientific study as personally relevant and meaningful requires that you show genuine curiosity about your students'

Connecting science activities to the students' lived experiences is one way to fully engage students in the process of scientific inquiry. Here, a cooking experience can become a science experiment.

Lawrence Migdale

Modifying biases

THE VOICE OF THE SCIENTIFIC AUTHORITY IS LIKE THE MALE VOICE-OVER IN COMMERCIALS, A DISEMBODIED KNOWLEDGE THAT CANNOT BE QUESTIONED, WHOSE AUTHOR IS INACCESSIBLE.

—*Elizabeth Fee*

interests. Look again at the list of "Questions to Ask About Your Students" in Chapter 1 (page 00). These questions, if you apply them conscientiously, can help you gain a rich insight into the lives that your students have lived before they arrive at your classroom door.

Do remember that all students come to school with a set of biases and beliefs about their world that are constructed from the social, cultural, and gendered contexts in which they live. These beliefs strongly influence students' tacitly held attitudes toward science and scientists. Part of your work as a science educator is to learn about your students' views of science and then use that knowledge to help them modify and enlarge their views.

As we saw in Chapter 2, science is frequently perceived as a white, male domain from which women and all people of color are absent. In addition, science is often portrayed in the mass media as a relatively impersonal field. For example, many television documentaries show the results of scientific work—the content of science—rather than the process of science or its human side. Science then seems a dry collection of facts and principles. This absence of the human face helps create in students a deep and abiding lack of interest in scientific study.

Hence, it is imperative for teachers to use science experiences to make connections with their students' daily, lived experiences. You can do this by relating scientific concepts to the natural events and artifacts of nature that comprise the students' own environment. In this chapter, you will see a number of ways in which teachers in ordinary classrooms integrate science experiences into the daily lives of their students.

SCIENCE CORNERS

One way to make science part of the daily life of the classroom is to reserve a table in a corner of the room that becomes labeled the "science corner" or the "science table." Sometimes the area is called the "science center" because it becomes the center of scientific activity in the classroom. Whatever name you choose, the science corner can reflect many possible activities in the life of an elementary school class. Here are a few ideas:

Ideas for science corners

■ *Seasonal changes.* The science corner can reflect ongoing changes in the seasons. For example, an autumn science corner may have acorns, pine cones, colorful leaves, chestnuts, and other tree parts that commonly fall to the ground at this time of year in many parts of the country.

■ *Materials related to the current science unit.* A science corner could reflect the science unit the class is engaged in at the moment. For example, a fifth-grade class studying ecosystems may create some small habitats in its science corner. A third-grade unit on sound energy may be reflected by student-made instruments in the science corner.

■ *A class nature collection.* A science corner can include a potpourri of items from nature brought in by students simply because the students found them fascinating. Such a science corner may be appropriately titled "The Class Collection from Nature."

Making the Science Corner an Interactive Experience

In many classrooms you can find objects of nature on display in a science corner. Unfortunately, if it is just a display area, teachers and students often lose interest in it. These objects merely collect dust, serving no useful

An invitation to explore

purpose. To make science connections with the students' lives, the corner needs an invitation to explore, a reason for the children to visit. The following sections offer some clues for turning the science corner into an interactive experience that engages students in science process skills.

Designing the Science Corner with Your Students   One way to interest students in the science corner is to involve them in designing it and supplying its contents. This participation will help ensure that the informal science curriculum reflects *their* interests.

Questions to ask

In setting up a science corner with your students, here are some questions you can ask that will engage them in the process:

What should we put in our science corner?

What can we *do* there?

What materials will we need when we explore the objects in the corner?

Can we keep a small class pet there?

What rules do we need about how and when to use the science corner?

Science corner crews

In the upper elementary school grades, students can largely control the contents and activities in the science corner. Science corner "crews" can rotate monthly, taking the responsibility of changing some displays and maintaining others, as they see fit. This regular attention will help ensure a regular flow of materials into and out of the corner.

For very young students, the teacher can take a more active role in designing the exhibits. Even in the younger grades, however, you can engage students in designing the science corner by offering possibilities and inviting them to share their own special objects from nature.

Questions and Activities   Like the science circus stations described in Chapter 5, a classroom science corner should have cards (or displayed pieces of paper) with questions. For example, if the science corner has a collection of acorns, the card could read, "How would you classify these acorns? Why did you classify them this way instead of some other way?" Whenever possible, the students themselves should construct the questions when they set up the displays. The students often answer the science corner questions in their science journals.

Encouraging students to experiment

In addition, the questions should encourage the students to explore and experiment with the materials at hand. In one second-grade science center, there were eyedroppers and a cup of water. Next to this set-up were some

Guidelines for Creating a Science Corner in Your Classroom

- The classroom science corner should reflect the students' interests and, wherever possible, be set up by the students.

- The science corner should be in a part of the room that is easily accessible for all students.

- For students with visual impairments, all signs and cards at the science corner should be in large, bold lettering and easy to read. Similarly, for students with physical disabilities, the signs should not be placed too high or too low for them to read easily.

- No objects that could pose a danger to children should be kept in the science corner—for example, no heating devices or objects with sharp edges.

- The corner should include tools for measuring different properties of objects, such as a balance scale, uniform masses for weighing, centimeter sticks for measuring distance, and graduated cylinders for measuring liquid volume.

- The center can also include other common science tools, such as magnifying lenses, prisms, magnets, and mirrors. The questions and activities should give students reasons for using these tools.

- If the classroom has a computer, it should be in the science corner. It should have some meaningful software programs to engage students in problem solving, and the questions posed in the corner should often lead students to use the computer as a tool. If the classroom has more than one computer, then at least one of them should be in the science corner.

- If the science corner reflects the theme of the current science unit, set out materials and experiences that extend the students' thinking. For example, for an electricity unit, set up the science corner so that it invites students to construct different kinds of circuits. The corner might have books about electricity and perhaps some small electrical appliances (unplugged!) for students to take apart and explore.

- The contents of the corner should change regularly, to keep stimulating students with new objects to investigate and new questions to ponder.

pieces of an old newspaper. The card read, "Place a drop of water on the letters in the newsprint. What happens?"

Imagine a science corner or science table that changes every few weeks, has an assortment of objects collected by the students and teacher, and challenges the students to find answers to questions. This is the sort of informal learning experience in which children can become deeply engaged.

Connections to Literature and Other Subjects

Science corners are perfect places to display the students' favorite science trade books or books of fiction that have some topical connection to the science objects on display. For example, one third-grade class I visited was studying local trees in the fall. The science corner, besides including many parts of trees to explore, had a *Field Guide to Trees* as well as Shel Silverstein's *The Giving Tree* and Dr. Seuss's *The Lorax.* (See Figure 6.1.) Although the latter two books are fiction, both connect important factual information about trees to the story line.

When the science corner includes books of fiction, it is often useful to ask the students to compare the story to what they know to be true about the subject. With *The Giving Tree,* for example, one of the cards at the science corner might ask, "In what ways is this story like what we know to be true about trees? In what ways is it different?" Even very young children will know—and enjoy explaining—that trees don't talk!

Books in the science corner

FIGURE 6.1
· ·

A science corner for third-grade students who are studying local trees. Notice the various materials to explore, the instruments for studying them, and the books that are kept handy.

Besides literature, science corners can link readily with other elementary school subjects. *The Lorax*, for instance, has important implications for the environment. Set in silly rhyme, it tells a powerful story of the results of constant logging. With this book in the science corner, you could connect a science unit on trees to social issues as well as to literature. Integrating mathematics also becomes routine as we do science with children. In one third-grade class, the science corner displayed a cross-section of an old tree trunk and invited the children to count the tree rings and predict the age of the tree. Remember how the water-drip station in the science circus in Chapter 5 (p. 94) involved students in measurement and calculation.

As we continue through Part Two of this book, you will see many possible links between science topics and other subjects. By taking advantage of these connections, you can strengthen the interdisciplinary nature of the elementary school learning experience, while at the same time developing your students' abilities to make their own connections.

Sample Science Corners from Around the Country

Looking at some science corners that teachers and students have set up in different areas of the country will stimulate your thinking about ways to connect science concepts with your students' daily lives.

A Third-Grade Science Corner in the Northeast   This science corner has a collection of seashells in a plastic container. An index card on the table next to the shells has neatly printed on it:

> Can you classify these seashells? In what ways are they the same? In what ways are they different?

A Second-Grade Science Corner in the Northwest One second-grade student has brought a very large pine cone to class. It becomes part of the science corner. The teacher asks the students what they want to find out about this pine cone. The children wonder how much it weighs, where it came from, and why it is so big. As the pine cone is placed on the science table, the teacher writes on the accompanying question card:

How many red chips weigh the same as the pine cone? How long is the pine cone? Where do you think it came from? Why is it so big?

The children who wander over to the science corner are free to choose any or all of the questions to explore.

A First-Grade Science Corner in the Midwest A stuffed, furry yellow chick sits on the science table. The index card reads:

In what way is this baby chick like a real chick? In what way is it different?

A Third-Grade Science Corner in the Autumn   A group of leaves sits on the table. They have different shapes, sizes, and colors. An index card reads:

What types of trees do you think these leaves came from?

A Second-Grade Science Corner   A small basin of soapy water is surrounded by pipe-cleaner shapes that look like bubble wands. Some are triangular, and some are square. The index card reads:

What shape is a bubble?

In the same science corner, next to a prism, an index card reads:

> Hold this up to the light. What colors do you see?

A Third-Grade Science Corner   A basin of water invites the students to place different objects in the basin to see if they sink or float. The objects on the table are a marble, a paper clip, a wooden block, and a small ball. The children are allowed to try other objects as well (but dunking peanut butter sandwiches from their lunch boxes is discouraged). An index card asks:

> Which objects sink in water? What happens when they sink?

FIELD TRIPS

Another common way to link science to the world outside the classroom is to take the students outside—on a short stroll around the school grounds or a more extensive field trip.

I was visiting a northeastern elementary school one winter morning when a teacher I knew said, "Hi. The class and I are going to visit the Hall of Science today." This is a fine hands-on museum not far from the elementary school. I responded, "Great! How come you are going now?" She said that she had planned the trip weeks ago and thought that, at this time of year, it would be good to get the children out of the classroom.

Making field trips meaningful

It struck me that something was lacking in her definition of the trip's purpose. Certainly it is always a treat, if your school has the resources, to take students on a class trip, but science field trips can be much more meaningful if they are directly connected to the unit or topic the class is

Extending the science experience beyond the classroom enriches learning and makes connections to the school's local environment. Here, students learn about reptiles while on a field trip.

E. Williamson/The Picture Cube

Tips for Outdoor Science Experiences

- Careful planning is essential. For field trips, letters of consent are usually required from parents or guardians, and students should be guided to dress properly for the chosen site.

- Remember to prepare the materials you need to take to collect samples and record information. Ziploc bags, plastic collecting cups, student science journals and pencils, and magnifying lenses will always come in handy in the great outdoors.

- Students should understand the reason for the outdoor trip and have some ideas about the problem they are exploring.

- The trip should lead to further investigations or research in class, as illustrated by this science journal entry by a young student named Ana:

  I went on a field trip with my class to the Duck Pond. We were looking for something new we had never seen before. When we came back to the room we all guessed what it was. Based on our past experience we all said it was either a bee hive or a wasp's nest. We tried different experiments to find out what it was. After researching it we identified it as a praying mantis home. (Reprinted by permission.)

- Adult supervision is required. Depending on the type of excursion and the number of students you are taking, you may need to ask other adults to accompany you. These adults should also be part of the group planning, if possible.

studying. By linking the trips with class study, we reinforce the learning experience—and we demonstrate that science is more than a subject confined to the classroom.

Exercising process skills

In addition to relevant science content, a field trip should provide students with the opportunity to exercise their science process skills. Invite your students to engage in some problem solving outdoors. For example, a fourth-grade trip to the seashore might involve taking notes, collecting articles of interest, and perhaps using a ball or two to explore the actions of the ocean waves. Students could be engaged in recording their observations, drawing what they see, and solving one or more problems that they have brainstormed previously in class.

Questions in planning a trip

Whenever possible, share your rationale for venturing outdoors with the students. Involve them in planning both the excursion itself and the questions it will seek to answer. Ask them, for instance:

Why do you think we are going there?

What shall we bring?

What do you wonder about?

Your students will have good ideas about the trip, plus their own stories to tell about related experiences.

In the following stories, we will see some examples of informal science learning experiences both in and out of the classroom.

SCIENCE STORY

Making Connections, Inside and Outside the Classroom

The Osage

When a fourth-grade student in southern New York State brings an interesting-looking fruit to class, neither the teacher nor the other students have ever seen anything like it.

"This fell from a tree near my house, and I don't know what it is. Can we put it in our science corner?" the student asks.

"That's a very good idea," remarks the teacher. "What question can we put on our index card?"

"Let's ask, 'What do you think this is?'" the student replies.

Over the next few days, the class becomes intrigued by the strange-looking fruit. The children write in their science journals the qualities they notice about it. It is green, they observe, with thick, bumpy skin like an orange. It also smells a little like an orange. They wonder why the skin is so thick.

What is this strange fruit?

After recording their observations and questions, the students explore the school library and finally ask for assistance. With the librarian's help, they find a picture and description of the fruit in a resource book. It is an osage, they learn. But their curiosity has been piqued, and they don't stop there.

Now they read about where the tree grows, how it got its name, and much more. They learn that an osage is the fruit of the osage orange tree, which belongs to the mulberry family of trees. It was originally found in Texas, Oklahoma, and Arkansas and named for the Osage Indians of that

Where is it from?

region. It is an inedible fruit, and its tree has a short trunk and crooked branches. It is planted around the United States for hedges and ornamental purposes.

For the rest of the school year, everyone in that fourth-grade class looks for osage orange trees.

A Snow Story

One winter morning after a heavy snowfall, a fourth-grade teacher in Nebraska uses an empty coffee can as a scoop and fills it with snow. She asks the students what they can learn from it. She knows that they have seen the snow every winter and probably have many unanswered questions about it.

"What are your questions?" she asks. "What do you wonder about? How can you find out?"

The students wonder if all snowflakes are really different. Why is it so quiet when it snows? Is snow a solid, liquid, gas, or all three? When the snow melts, will it still fill the can? What temperature is snow?

Questions about snow

They take the can of snow to their science corner, where they measure its temperature with a thermometer. They examine small bits of the snow under a microscope and discuss what they see. They predict how high the water level in the can will be when the snow melts. They are shocked to see what a tiny amount of water the can full of snow melts into! They wonder what takes up all the space in the snowflakes. Yes; it is air, and it is the air spaces between the snowflakes that absorb sound and make it seem so quiet on a snowy day.

A surprising discovery

On that day, the snow becomes the science curriculum. The formal curriculum waits for a while. The snow will not last forever, and the moment is there for the taking.

Two Seashores

With collection bags in hand, children in a fourth-grade class on Long Island visit the south shore, the ocean side of the island. The sand forms smooth dunes here, and its fine grains glisten in the autumn sun. The students draw pictures of the sparse plant life on the beach, and they study the shells washed up. They collect various artifacts to take to their science corner. Back in the classroom, using reference books, they try to identify the marine invertebrates whose shells they have found.

Some weeks later, they visit the north side of the island, where the bay gently brushes against a craggy shore. They notice how very rocky and pebbly it is. Encouraged by their teacher, the children begin to wonder

Why are seashores different?

Exploring a Coastal Plain

As teachers, we often have fascinating natural phenomena right under our noses, but familiarity tends to breed indifference. The first step in bringing the local environment alive for our students is to pay attention to it ourselves.

In a small coastal fishing community in Maine, ten elementary school teachers accompanied me to a part of their coastline, an inlet where granitic mounds of rock dotted the landscape. This place was just down the road from the elementary school, and we were discussing how to use their local environment as a living laboratory for their students. For me, it was a mysterious and interesting site, but for these local teachers it seemed quite ordinary because they had seen it countless times before.

"What do you notice?" I asked them. "Tell me everything, anything." Thus prodded, they began to observe more than they had suspected they would.

For example, it was low tide, and my companions pointed out that all the mounds of rock had water marks, indicating the point to which high tide had risen that day. *Tides*, these teachers knew, are the rise and fall of the oceans caused mainly by the moon's gravitational pull on the earth. The waters of the earth on the side of the earth facing the moon bulge out from the moon's gravitational pull, creating a **high tide.** On the other side of the earth, there is also a high tide, because the moon's gravitational pull has pulled on the solid earth as well, leaving the waters bulging on the opposite side of the earth. Meanwhile, as the waters bulge on two sides of the earth, the remaining waters flatten, causing lower levels of water, called **low tides.** Because the earth rotates on its axis once every 24 hours, the tide rises for about 6 hours at a particular point, then falls for about 6 hours.

The teachers knew all this, but it hadn't occurred to them that they could use the inlet down the road from their school as a place for their students to investigate the concepts of tides, gravitation, and the movements of the earth and moon.

As we stood there, we noticed that some untidy seagulls had left the shells from the mussels they had eaten atop a large rock, which had served as the gulls' dining table. That observation sparked a discussion about food chains.

One teacher picked up a dried scallop shell in which a spider had spun its web. Talking about it, we brought up ideas about animal habitats, ecosystems, and more.

These teachers were so impressed with the natural phenomena under their noses that they took their students to the inlet the next day to collect materials for science corners.

why this north shore looks so different from the south. They make drawings and return with sand and rock samples to place in their science corner for comparison with the samples from the south shore.

In the next weeks, using books, CD-ROMs, and the Internet, they learn about the effects of the ocean and inlet bays on landform. They make inferences, come up with their own ideas, and research areas of personal interest. The comparison of the two shores yields some important science concepts.

Investigating a Natural Disaster

A major earthquake strikes California. News of the sudden destruction floods the television and print media nationwide. In Texas the next day, one fourth-grade teacher begins doing "earthquake science," despite the fact that it is not listed in the formal fourth-grade curriculum.

The students clip newspaper stories of the disaster and hang them around the room. Prompted by the teacher, they write down their earthquake questions and research them in library books and on a web site. In the science corner, some students make clay models of the earth's crust and use plastic knives to cut lines in the clay representing faults. Others use layer cakes in cake pans that are precut before the cakes are baked. (When you bake a layer cake in a pan that has been cut slightly on the bottom, the layer comes out with a large crack, resembling a fault in the earth's crust.) With these models, the students learn that an earthquake is a sudden slippage of rock that sends enormous shock waves through the earth.

Making models to understand an earthquake

The students also write stories about the earthquake and take collections for the earthquake victims. Besides increasing their understanding of earthquakes in particular, the explorations help make the children aware of the impact of natural events on human lives.

A Tree Grows in the Bronx

In the Bronx, a borough of New York City, one elementary school occupies a full city block. Overcrowded and understaffed, it has several different shifts for student lunches. The local community, largely African American and Latino, is rich in energy but not in resources. The school yard is fenced in. On nice days, the students can play games in the yard, but they are not allowed beyond the chain-link fence. Their daily environment is intensely urban.

These children have little exposure to trees, fields, wildflowers, streams—the kinds of settings we commonly think of as "nature." Yet an old oak tree grows in the corner of the school yard, where nothing else

THE COMMITMENT TO SCIENCE FOR ALL IMPLIES INCLUSION OF THOSE WHO TRADITIONALLY HAVE NOT RECEIVED ENCOURAGEMENT AND OPPORTUNITY TO PURSUE SCIENCE.

—National Science Education Standards

seems to live. A first-grade teacher notices this tree and seizes on it as a way to explore the changing seasons with her students.

The children visit the tree in the early fall. Its leaves are still green, and they notice the acorns attached to its branches. They draw what the tree looks like. After that, they visit the tree weekly. Later in the fall, they see the leaves change color and the acorns drop. They take some red leaves and acorns inside to display in their science corner, where they have a tree book and a magnifier. They identify the type of tree and learn some ways in which it differs from other trees.

Finding nature in the city

In the winter, they make drawings of the tree's barren appearance. They watch as snow accumulates on the branches. In the spring, they notice as the sun prods the tree to develop new leaf buds.

By the end of the year, their drawings of the tree in every season fill an entire wall of their room. In the midst of the concrete maze of buildings and school yard, they have used a lone red oak tree as a way to explore seasons.

E X P A N D I N G   M E A N I N G S

The Teaching Ideas Behind These Stories

■ Science comes alive when you think of it as a way to explore nature in your own backyard. In each of our stories, the students' scientific inquiry was connected to where they lived.

■ We can say that these students followed a "both/and" curriculum (McIntosh, 1983). That is, they pursued both the prescribed, formal curriculum of the school and their own informal, self-constructed, experientially grounded, curriculum. In effect, an informal curriculum constructs a new textbook—the textbook of our daily lives. ("Making textbooks of our lives" is a phrase first used by Emily Style, codirector of the National SEED (Seeking Educational Equity and Diversity) Project on inclusive curriculum, based at the Wellesley College Center for Research on Women, Wellesley, Massachusetts.)

■ Connecting science to students' own lives—in a way that encourages them to explore their own questions—both engages them in using science process skills and validates their sense of themselves as scientific thinkers. This is particularly important for students who might be discouraged or marginalized by traditional science teaching.

What Kind of Land Are You On?

The geologic diversity of the United States offers wonderful opportunities for students to explore topics in earth science. Typical kinds of landform include the following:

- *Mountains* are great masses of rock pushed high into the air by forces from deep within the earth.

- *Plateaus* are flat surfaces that are high in elevation compared to the land around them. Extensive plateau areas are found in New Mexico, Arizona, North and South Dakota, Montana, and Wyoming.

- *Peneplains* (literally, "almost-plains") are flat surfaces formed as old mountains wore down until they were almost level. Southern New England is an example of a peneplain.

- *Plains* are flat surfaces that are low in elevation compared to the land around them.

- *Coastal plains* are made up of pieces of rock that were worn away from rocks along the seashore by ocean waves or carried to the ocean by rivers.

- *Glacial plains* are made up of rocks deposited by a glacier that spread over the area.

- *Interior plains* were formed from large, shallow, inland seas when sediments filled these seas and the water disappeared. The Great Plains of the United States is a region of interior plains.

- *Lake plains* were formed when an old lake floor was lifted up by geologic forces or when climate conditions caused the water to drain away. Minnesota and North Dakota are made up largely of lake plains.

- Since classroom science corners should reflect the lives of the students, it makes sense that they vary with the geographic location of the school and the interests and backgrounds of the students. For instance, classrooms in coastal regions may be rich in seashell collections, and inland regions may have rich rock collections. A southwestern school may have a marvelous display of different sands and sedimentary rocks on a science table. This variation is appropriate and important since it helps students make meaning of their local environment.

- As we saw in Chapter 4 and again in the snow story in this chapter, changes in the weather offer valuable opportunities to connect science to students' lives. Similarly, local or national news often has important scientific implications, as in our earthquake story.

The Science Ideas Behind These Stories

- Osage trees are distinctive because of their thick, pebbly skin. Like many other plants that originated in dry climates, they have evolved an outer coating that protects the fruit from dehydrating. They are not common in southern New York State.

- Snow forms from water vapor that condenses when the temperature of the air is below freezing. The term *condenses* means that the particles of water vapor come close together, forming droplets of water. When the temperature is below the freezing point, the water vapor condenses into snow crystals rather than into rain. Snow, then, is frozen water vapor. There is a lot of space between snowflakes. The air fills the spaces between the snowflakes, and that is why snow takes up so much more space than the water it turns into when it melts. That is also why the environment seems so quiet after a snowstorm; sounds are absorbed by the air spaces.

- As the students in the Long Island story discovered, seashores have different-size particles, depending on how the shore was created. On the south shore of Long Island, the constant wave action of the Atlantic Ocean causes rocks to be worn down into tiny sand particles—a process called **weathering.** On the north shore, the water is from a bay, an inland waterway. Here the water action is gentler, the process of weathering is much slower, and hence the rock particles are much larger and coarser than the fine-grained sand on the south shore.

- An earthquake is a sudden slippage of rock. Earthquakes occur because the outer layer of the earth (the part we live on), called the *crust*, is made up of huge chunks of rock called *plates*. Rock plates lying next to each other are pressed tightly together. Sometimes the pressure is so great that they need to slip past each other ever so slightly to settle in a new position that eases the pressure. This slippage is what we experience as an earthquake. Earthquakes are most common in areas where there are boundaries between the plates or huge cracks in plates. These boundaries and cracks are called **faults.**

- As the students in the Bronx observed from their lone oak tree, most plants in temperate regions grow only in the warm weather of the spring, summer, and fall. Their oak tree, which lost its leaves in the fall and remained inactive during the winter, is an example of a **deciduous tree** (a tree that sheds its leaves at the end of the growing season). Evergreen trees, also called **coniferous trees,** do not lose their leaves in the fall, but they too are inactive in the winter.

Questions for Further Exploration

- On what kind of landform do you live?

- How does nature present itself in your surroundings? Are you near a seashore, a desert, a grassy plain, a park rich with conifers?

- How many distinct seasons are noticeable in your area?

- In what ways have people changed the natural landscape of your environment?

- What is the local plant and animal life like?

- How can aspects of your local landscape inform your science corner?

- What formal science units are your children studying that can find their way to your science corner?

Resources for Further Exploration

Electronic Resources

Frank Potter's Science Gems—Earth Science II. **http://www-sci.lib.uci.edu/ SEP/earth2.html.** This page of Frank Potter's Science Gems web site has links to many resources for lessons on the atmosphere, weather, earthquakes, water, and more, grouped by approximate grade level.

Science Adventures. **http://www.scienceadventures.org.** This online directory makes it easy to find locations near your home that offer hands-on experiences, exhibits, and teacher and public educational programs. Search features include location within a specified radius of a zip code. The list includes zoos, parks, gardens, aquariums, nature centers, and planetariums.

Science Learning Network. **http://www.sln.org.** This web site has important links to museum web sites from all over the country and to other resources that make science connections to daily life.

USGS Earthquake Information. **http://quake.wr.usgs.gov**. Up-to-date information on earthquakes around the world, with maps.

Print Resources

Brainard, A., & Wrubel, D. (1993). *Literature-Based Science Activities: An Integrated Approach.* New York: Scholastic Professional Books. Designed for students in kindergarten through grade 3, this volume highlights children's books for introducing, extending, or enriching the science content in twenty topic areas.

Burnie, D. (1988). *Tree.* New York: Knopf. Photographs display all the varieties of coniferous and deciduous trees.

Gardner, R., & Webster, D. (1987). *Science in Your Backyard.* New York: Simon & Schuster. Encourages students in grades 4 through 6 to use their backyards, parks, and playgrounds as sites for scientific investigations.

Parker, S. (1989). *Seashore.* New York: Knopf. Using actual photographs, characteristic of this book series, this book provides a wonderful seashell reference.

Roth, C., Cervoni, C., Wellnitz, T., & Arms, E. (1991). *Beyond the Classroom: Exploration of Schoolground and Backyard.* Lincoln, MA: Massachusetts Audubon Society. How to use your school's immediate environment to create a familiar laboratory for activities in kindergarten through grade 6.

Trolley, K. (1994). *The Art and Science Connection: Hands-on Activities for Intermediate Students* and *The Art and Science Connection: Hands-on Activities for Primary Students.* Menlo Park, CA: Addison-Wesley. Sourcebooks of creative art activities that integrate art and science for students in kindergarten through grade 6.

Zolotow, C. (1992). *The Seashore Book.* New York: HarperCollins. A fictional story in which a mother's words help a little boy imagine the sights and sounds of the seashore even though he has never seen the ocean; appropriate for grades 2 through 6.

THE DAILY LIFE OF THE CLASSROOM

When you enter an elementary school classroom, the life of the class is often reflected by the materials adorning the room. I often look for a science corner. When I find it, I spend time exploring the materials. Sometimes they tell me what science unit the students are studying; at other times I learn about the students' own collections and interests. Often I learn something significant about the class from its science corner.

The tacit message I receive from classrooms that have scientific materials available to the students is, "This class engages in scientific activities. Science is a regular part of the life of this class." A good science corner, of course, does more than simply present materials that have something to do with science. It also invites students to explore science connections that are meaningful to them. It links whenever possible to local weather, news stories, and features of the surrounding environment.

Challenges and new ideas

The science corner can be a place where children are challenged to consider some questions and then generate their own questions. After visiting the classroom science corner, children frequently write in their science journals or notebooks, reflecting on what they did and what questions their experience raised for them. They also generate their own ideas for materials to add to the science corner, and questions to ask about these new objects.

Benefits of the informal curriculum

Similar principles apply to field trips and other outdoor activities, such as studying a tree in the school yard. By constructing an informal curriculum from natural items and events in the children's own world, we make it possible for them to value themselves as scientists and to overcome the traditional factors that alienate many young people from science. We help them understand science content more deeply and more personally, as part of a larger picture. Finally, we raise their awareness of their natural environment and the issues it presents for their future life as adults.

Key Terms

formal science curriculum *(p. 106)*
informal science curriculum *(p. 106)*
high tide *(p. 122)*
low tide *(p. 122)*
weathering *(p. 125)*
faults *(p. 126)*
deciduous trees *(p. 126)*
coniferous trees *(p. 126)*

7 Science Is Not Neat: Explorations of Matter

FOCUSING QUESTIONS

- What is important about the process of classifying?

- What types of things do you classify in your daily life?

- What types of "messy" explorations do you do?

- What do you think is meant by the statement, "Science is not neat"?

One of my favorite definitions of "science" refers to it as a way of making sense of the world. To me, that suggests in part that science helps us deal with the enormous diversity of the materials—both living and nonliving—that exist in our environment. In this chapter, we are going to look at classrooms where children are exploring nonliving materials and trying to make sense of them. This sense making usually begins with the process of **classifying**—the grouping of materials with similar properties into the same category. As you read the stories in this chapter, look for these themes:

- Classifying encourages students to make careful observations.

- Knowing *why* an object belongs in a particular category is as important as knowing *where* to place it.

- Choosing a category for an object or an idea isn't always simple.

- Science can be messy—in more than one way. Often you make a mess when exploring materials. Often, too, the ideas that are generated in the process of classifying are themselves complex and "messy."

CLASSIFYING

From simple
categories to
hierarchies

Using property words

As we discovered in Chapter 5, classifying is one of the basic process skills. For children, classifying begins with the simple recognition of properties and matures into a grouping and sorting process. The categories that very young children employ are usually mutually exclusive and not hierarchical (Siegler, 1991; National Research Council, 1996, p. 128). This means that the children are grouping and sorting into categories without thinking of a larger picture in which those categories themselves could be placed. The larger picture emerges as the students mature. By fourth grade, they are Toften able to classify an object into more than one category, and they are more and more able to arrange categories into hierarchies. Figure 7.1 offers a simplified schematic to illustrate this maturation.

In elementary school science, children often learn classifying skills by identifying the properties of materials through careful observation, and then comparing and contrasting those properties. The basic words we use to describe the material world are sometimes called **property words.** These are words that refer to common properties of objects, such as size, shape, color, odor, texture, taste, composition, and hardness. *Big,* for example is a property word, as is *red.* In developing their classifying skills through the use of property words, children are also learning language and communication skills.

In Chapter 6, we saw that science corners often have collections of objects from nature: seeds, flowers, seashells, vegetables, fruits, rocks, and fossils, to name a few. The activity most often associated with collections is classifying. For example, how are these seeds the same? How are they different? In answering such questions, children exercise all their senses (except the sense of taste, which teachers generally discourage them from using on classroom materials), and they learn how to apply many different property words.

In the following two science stories, we'll see how some students in the early grades identified certain properties as they attempted to classify objects according to their state of matter. The states of matter—solid, liquid, and gas—sound simple, but as we saw in Chapter 3, they are not always obvious or visible.

FIGURE 7.1 **Growth in children's classification abilities.** (a) At a young age, children can do a simple sorting of objects into mutually exclusive categories: soda is different from juice, and both are different from brick. (b) Later, the children recognize that the objects can be grouped according to their states of matter—in this case, solids and liquids. (c) Before they finish elementary school, children may understand that these (but not all) solids and liquids are part of a chemical class known as mixtures.

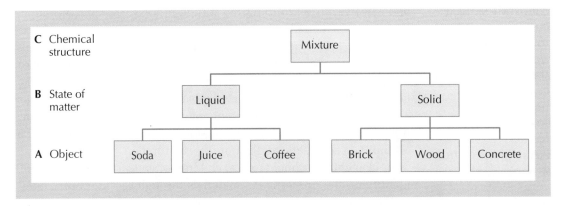

SCIENCE STORY

Exploring Solids, Liquids, and Gases

Ms. Harrison's multiage class is in a rural setting, far from the action and resources of metropolitan areas. When I enter, I see that she has set up five stations around the room so that the students can work simultaneously in groups. It is late in the school year, and the children have learned the rules of their classroom setting. They seem comfortable and are functioning well independently. The class has nineteen students at different stages of development, in grades 2 and 3.

Ms. Harrison has prepared five identical sets of materials, one set for each group of students. Each set has three plastic bags, all the same size. In one bag she has placed a wooden block; another bag holds 8 ounces (1 cup) of water; and the third bag she has blown full of air. All the bags are closed with twist ties.

She plans for the children, in groups of four, to explore these bags and describe their contents. To introduce the activity, she says, "We have the opportunity to explore some materials today. We are going to observe them in any way we can, and then we will try to classify them. In your science journals, write down all the things you can say about these materials."

The only labels on the bags are as follows: All the bags with solid blocks in them are labeled "bag 1." All the bags of water are "bag 2," and all the bags of air are "bag 3."

Why such vague labels?

Ms. Harrison's thinking: Although Ms. Harrison is interested in having the children recognize the differences among solids, liquids, and gases, she sees no reason to supply those labels up front. She wants the children to do their own thinking about the classifications, and the labels can be introduced later.

Ms. Harrison guides the children along: "What are all the things we can say about the materials in bags 1, 2, and 3? Remember to write your observations in your journal."

Suddenly, Alena, one of the older students, calls out, "That's easy, Ms. Harrison. One is a solid, one's a liquid, and one's a gas!"

Alena supplies the names.

"Yes, Alena, that is one type of label we can give these materials, but I asked all of you to make *observations*. Let's talk about *why* one is called a solid, while another one is called a liquid and another one a gas. Your observations can help us with that."

Ms. Harrison's thinking: Okay, Ms. Harrison thinks, the labels are more up front than she planned. Several children already know that these materials are ex-

Weighing and measuring are ways to learn about the properties of different types of matter. The students here are using a simple double-pan balance and weighing different objects.

Martin Miller/Positive Images

amples of solids, liquids, and gases. But those are just labels. She is hoping for some meaning behind the labels. She wonders whether the students will be able to determine what properties are usual for any solid, liquid, or gas by exploring these materials.

As the students examine the three bags and write their observations in their journals, Ms. Harrison visits each group. Suddenly there is a scream. One of the water bags has spilled all over the table and Todd's leg. Another student gets paper towels, and the group cleans up the mess. Todd says, "Sorry!" It appears that the bag broke because he handled it a bit too much, or too roughly.

An accident!

After making sure that Todd does not mind a wet pant leg, Ms. Harrison says, "Sometimes science can get messy, but that's okay. Why don't you put some water in another bag, Todd. Tie it tightly, and go on with your observations."

My thinking: As I watch the class working, I am impressed with the calm way Ms. Harrison handles the spilled water. Certainly this is not the first time a spill has happened, and it won't be the last. I agree with what she told the class—that getting messy is sometimes a part of doing science.

When the children appear to be finished writing in their journals, Ms. Harrison gathers their observations and records them on a class chart. Here are some of the children's responses:

The children's observations

The block	The water	The air
It is yellow.	It is white.	It is light.
It is hard.	It is wet.	I can see through it.
It's a rectangle.	It is drippy.	I can tap it and make it fly.
It has 8 corners.	It has no shape.	It has no shape.
It has 6 sides.	I can see through it.	It doesn't drip.
It can float in water.	I can pour it into a glass.	
I can bang it on the table. It makes a loud sound.	I can squeeze the bag.	
It doesn't change its shape.		
I can't squeeze it.		
It's smooth.		
It is medium sized.		

My thinking: I see that the distinctions among the states of matter have begun to emerge in the students' minds. Now the lesson is really going beyond the labels.

Ms. Harrison begins to pull the lesson together by remarking, "Alena said, when we began, that these materials in our plastic bags were a solid, a liquid, and a gas. What do you think about that?"

After discussing the question, the students conclude that the air is an example of a gas, the block of a solid, and the water of a liquid. Ms. Harrison tells them these are three states of matter. Scientists say that matter is anything that has weight and takes up space, she goes on. She urges the students to see if their three samples meet those criteria. Do they all take up space and have weight?

The children agree that the block, the water, and the air take up space, but not everyone is convinced that the air has weight.

Does air have weight?

"How many of you think that we can weigh the air?" Ms. Harrison asks. About half the students raise their hands.

Ms. Harrison proceeds to blow air gently into another plastic bag and tie it. She rests a book on the bag of air and invites the children to observe how the air takes up space and supports the weight of the book. Then she asks, "How can we prove this air weighs something?"

My thinking: As I know from other experiences—like the ones described in earlier chapters—the gaseous state of matter is least concrete for children and most difficult for them to grasp. I think that Ms. Harrison is wise to pursue the idea that air can be weighed.

The children aren't sure how to weigh the air. Ms. Harrison takes a double-pan balance and places an inflated balloon on one side of the balance and a deflated balloon on the other side. Before she lets go of the balloons in the pans, she asks the children what they think. How will the scales tip? The children are divided about the outcome.

Asking for predictions

The inflated balloon tips the scale downward, indicating that the balloon filled with air weighs more than the balloon with no air. Then Ms. Harrison asks the children to blow on their hands, feeling their own exhaled air.

"It tickles," Julia says. "It feels warm," remarks Marika.

Ms. Harrison's thinking: Ms. Harrison hopes that all these little "air" experiences will help strengthen the idea that air is matter: in addition to taking up space, it has weight, you can feel it, and it can be warm or cold.

Ms. Harrison continues, "Let's look at the list of properties on our chart. Notice that we had more things to say about the solid than about the liquid or gas. Why do you think that is so?"

My thinking: I see that Ms. Harrison is beginning to generalize from the block, the air, and the water to solids, liquids, and gases. It is true: solids have more attributes for children to describe than liquids and gases, and it is easier for children to come up with property words that apply to solids.

The children go over their list of the wooden block's properties. They begin to understand that solids have surfaces and shapes and textures, and that they are easy to measure compared to objects with changeable shapes. Ms. Harrison invites the students to name the solids in their classroom. The students generate a long list of objects: tables, chairs, walls, posters, books, and so forth.

Listing properties of solids

Ms. Harrison then asks, "What liquids do you drink or see every day?" The students generate a long list again. She says, "Let's look at our bags of water. How can we weigh the water?" The children ask her to put one bag on a scale, and she does so for them. She points out that the bag weighs something as well, but its weight will be too small to notice.

How can we weigh a liquid?

"What if we were weighing our water in a cup?" Ms. Harrison asks. Some of the older students remark that they would have to weigh the cup first, then weigh the cup and water together, and then subtract the weight of the cup.

My thinking: I notice how Ms. Harrison varies her tactics. She asks the students to find examples of solids in the classroom; but when she turns to liquids—realizing there will be few or none in the room that the children can see—she changes the question into one about their everyday lives. I also like the way she asks them to think about weighing the water in the cup, wanting them to understand that weighing the liquid alone is not that easy. (I am reminded of the students in Mr. Wilson's class who weighed the melting icicles.)

For the next day, Ms. Harrison asks the children to write down all the solids, liquids, and gases they notice in their own homes. And with each item, they should write *why* they think it is a solid, a liquid, or a gas.

Finding examples at home

My thinking: I'm reminded of the story Richard Feynman told (Chapter 3)—just knowing the label for an object doesn't tell us very much about the object itself.

Mysterious Matter

Ms. Hager, a first-grade teacher in a primary school in south Florida, is combining 2 cups of cornstarch with 1 cup of water in a large bowl. She is creating an activity that will be in some ways even more challenging than the one in Ms. Harrison's class. She is going to take advantage of the fact that it is sometimes difficult to decide what state of matter an object is in. Sometimes matter can have properties of more than one state.

Before mixing the ingredients, she has invited the children to explore the cornstarch and decide its state of matter. This generates a lot of discussion. The children think it is a powder, and they have a hard time deciding if a powder is a solid or a liquid. Ms. Hager challenges them to think about it but does not tell them the answer.

What is a powder?

"Where have you seen this white powder before?" Ms. Hager asks. One student says that her mother uses it to cook. Another student says that his parents use it to diaper his baby brother.

Now Ms. Hager takes 1 cup of water and adds green food coloring to it. Then she adds the water to the cornstarch and stirs the two together. This produces a peculiar green goo. After the students help her cover all the tables in the classroom with newspaper, she distributes a piece of the green goo to each pair of students, placing it in an aluminum foil pie pan. She encourages the students to explore the mixture, discuss it, and express all their possible observations.

A peculiar green goo

SIMPLY NOTICING THE DIVERSITY OF STUFF IN THE MATERIAL WORLD CAN EVOKE AN OPENMOUTHED SENSE OF WONDER AKIN TO VISITING A ZOO FILLED WITH ANIMALS YOU HAVE NEVER SEEN BEFORE.

—*Ivan Amato*

The students are delighted. As they quickly discover, the goo looks and feels like a liquid, but it can be shaped into a ball. It looks as if it should splash when they drop an object in it, but it does not. All in all, the "gushy mess," as one student describes it, provides them with a wonderfully tactile experience. As they get it all over the newspaper-protected tables, they come up with many observations about it. One student records in her journal that the mixture is "soft, smooth, and mushy." Another writes, "It was a little like a solid and a little like a liquid." Still another student observes, "We made green goo. If you hold it in your hand and squeeze, it drips!"

Experimenting with the "gushy mess"

Ms. Hager's thinking: Ms. Hager knows that it is important, especially with young children, to provide opportunities for them to explore interesting materials and use language to describe what they are experiencing. And she knows that this mixture will provide a good workout for their skills of observation and description.

"What happens when you add more water to it?" one girl asks.

"Let's try it!" replies Ms. Hager.

The students observe that adding water makes the mixture more like a liquid. Ms. Hager says that the measurements of the two ingredients are important.

Changing the proportions of ingredients

Another student wants to add more cornstarch. Again, they try the experiment, and it makes the mixture seem more like a solid. It also turns it a paler shade of green.

Ms. Hager challenges the children to decide what state of matter the green goo is in. They cannot make up their minds. They decide to place it in a new category for their matter unit: "Not Sure."

What's the answer?

Already they have a list on the board, and now they add their latest observation to it:

The children's categories

Solids	Liquids	Gases	Not Sure
wood	water	air	green goo
chalk	orange juice	soda bubbles	
blackboard	milk	helium	

They have learned that sometimes materials can fall into two categories—or none at all!

E X P A N D I N G   M E A N I N G S

The Teaching Ideas Behind These Stories

■ In Ms. Harrison's class, the important distinctions between solids, liquids, and gases *emerged* as the students examined the contents of each bag. Ms. Harrison did not rush to label the materials. Even Alena, the student who called out the correct labels, was missing the point. The main question is, "What does it *mean* when we say something is a solid or a liquid or a gas?"

■ When Todd broke the water bag, Ms. Harrison treated the mess as routine—a result of scientifically exploring water. It's important that students see exploration as nonthreatening. In an atmosphere where students are terrified to have an accident, honest explorations cannot occur.

- Notice how Ms. Harrison used several techniques to demonstrate that air is matter. This provided multiple opportunities for students to grasp the concept and relate it to their own experiences.

- When Ms. Hager asked the students where they had seen cornstarch before, she was making connections between the science experience and their experience outside the classroom.

- Ms. Hager did not tell her first graders the answer to the state of matter of the cornstarch. She let their dilemma sit with them for a while. As they worked with the cornstarch, they recognized that it must be a solid.

- Inviting young children to explore the cornstarch and water mixture was a way to encourage their use of descriptive language.

- Oobleck is the name sometimes used for the cornstarch and water mixture. It comes from the Dr. Seuss book *Bartholemew and the Oobleck* (1949). Whether or not you call the mixture oobleck, the Seuss book makes a great literature connection. So does *Horrible Harry and the Green Slime* (1989).

The Science Ideas Behind These Stories

- There are four **states of matter:** solids, liquids, gases, and plasma. For practical reasons, the first three are the ones we can explore with elementary school children. (Plasma, the fourth state of matter, is the random array of very hot matter—gases so hot that they no longer resemble any known gases because electrical particles have been stripped away from the central part of the molecules. Plasma is the stuff of which stars are made, and it is the most common state of matter in the universe, but not on earth. A tiny bit of plasma can be found in a fluorescent light bulb.)

- **Solids** have a definite shape and can hold that shape for an indefinite amount of time if outside conditions remain the same. Besides shape, children come to know that solids have a definite size, texture, and color and can be easily measured and weighed on a scale.

- **Liquids** have a definite size but no definite shape. Liquids take the shape of their container. The most commonly found liquid on earth is water. Many foods have a large water content—as does the human body.

- **Gases** have neither a definite size nor a definite shape. Gases take the size and shape of their container.

- Air is a mixture of gases. While oxygen is the most important part of the air to our bodies, only about 21 percent of the air is oxygen. Air is about 78 percent nitrogen.

- Cornstarch is a powdered solid. Even though it does not have rigid surfaces, the clump of powder stays together on a surface, in its own small mound.

- The cornstarch and water mixture is actually a **suspension;** the solid particles are literally suspended between water particles.

- Just why the cornstarch and water mixture has such odd properties, acting sometimes like a liquid and sometimes like a solid, remains a mystery to scientists. There are several theories about electrical charges and a theory about molecular size—but no conclusions.

- Frequently objects do not fall into neat categories. Understanding that the study of science does not always provide clear answers is an important way to make school science more like real science.

Questions for Further Exploration

- For the experiment that Ms. Harrison's class performed, what other materials might you use in place of the block of wood, the water, and the air in the plastic bags?

- Think about gelatin. What state of matter is it? What about mayonnaise? Oatmeal?

Resources for Further Exploration

Electronic Resources

ExploraNet. **http://www.exploratorium.edu.** This web site, sponsored by the Exploratorium science museum in San Francisco, is full of interesting articles and experiments to try, including an activity that involves Outrageous Ooze, which (did you guess?) is another name for oobleck or green goo.

The Four States of Matter. **http://demo-www.gat.com/SlideshowFolder/FourStates.html.** This site provides nice descriptions and illustrations of the states of matter.

Print Resources

Goldhaber, J. (1994, Fall). If we call it science, can we let the children play? *Childhood Education*, pp. 24–27. This article describes one teacher's dilemmas about getting "messy."

Sneider, C. (1988). *Oobleck: What Scientists Do. GEMS.* Berkeley, CA: Lawrence Hall of Science. GEMS stands for Great Expectations in Mathematics and Science, a curriculum writing project from the Lawrence Hall of Science. To order GEMS materials, call (510) 642-7771.

IF IT'S SO MESSY, CAN IT BE SCIENCE?

Two kinds of
messiness

In Ms. Harrison's and Ms. Hager's classes, the children were involved with what we might call messy investigations. They were messy in a literal sense: the water spilled, the green goo got all over things. In fact, if you try the cornstarch experiment, it is a good idea to cover the tables or desks as Ms. Hager did. Luckily, the mixture will wash off readily with plain water.

Second, there was a metaphorical messiness. The green goo behaved like a liquid sometimes and like a solid at other times. You may think that classifying objects, including living things, is a neat and clear process. It is not. As you will see in Chapter 8, there used to be two kingdoms of living things, and now we have five kingdoms. The more we know about things, the more categories we need; and sometimes, when our categories fail us, we end up with a puzzle, at least temporarily.

In the next sections we will look in more detail at these two types of messiness and examine their implications for doing science with children.

Messiness in the Classroom

Often new teachers find themselves making a mess with their students as they engage in science experiences, just as Ms. Hager and Ms. Harrison did. Should you worry about this?

Several years ago, while I was visiting a suburban elementary school in the Midwest, the principal said, "I can't wait to take you to our science room. We have a science teacher who is so wonderful that when you walk in her room, you can hear a pin drop."

"When the children are in there?" I asked.

Matter and Energy

The exploration of nonliving materials belongs to the branch of elementary school science called **physical science.** The study of physical science encompasses interactions between matter and energy.

As we saw in Chapter 4, **matter** means anything that has weight and takes up space. If you can measure it and weigh it, it is usually matter. By **energy,** we mean the ability to produce a change in matter. Scientists define energy as the ability to do work, and work is defined as a force moving through a distance. So, for example, energy is needed in order to push a rock off the edge of a cliff. The amount of force that is used and how far the rock moves could be a measure of how much energy is needed to move the rock.

Matter and energy are two very broad categories in which the diverse materials of the world can be placed. Matter can be further described by its different states—solid, liquid, gas, and plasma. By the upper elementary grades, students often begin to learn about chemical composition as well, distinguishing elements from compounds and mixtures.

An **element** is the building block of all matter. The simplest form of matter, it is made up of only one type of material. Examples of elements are iron, nickel, gold, silver, oxygen, hydrogen, helium, carbon, and mercury. There are ninety-two naturally occurring elements in the universe, and fourteen others that have been produced by scientists in laboratories. The smallest part of an element that still has the properties of that element is called an **atom.**

A **compound** is a combination of two or more elements in a definite proportion. For example, water is H_2O—two parts hydrogen to one part oxygen. The smallest part of a compound that still has the properties of that compound is a **molecule.** Molecules are made up of atoms. Salt, carbon dioxide, and water are three common compounds. When compounds form, the elements that comprise them lose their original properties.

Finally, a **mixture** is any combination of elements, compounds, and other mixtures. Because there are only ninety-two naturally occurring elements, most matter that we encounter is either a compound or a mixture.

Similarly, there are many forms of energy, and by the later elementary grades, students commonly begin to distinguish them. For example, what do you think of when you think of energy? Well, you may think of the energy you exhibit in running or playing tennis. That type of energy is the energy of movement. Scientists call the energy an object has because it is moving **kinetic energy.** A moving car, a falling rock, and a strong wind all have kinetic energy.

Objects can also have energy when they are not moving. The stored-up energy that your body has when you are sleeping the night before a big race is called **potential energy.** In most cases, potential energy derives from position in space. A rock about to fall off a very high cliff has greater potential energy than a rock about to fall off a low cliff. The first rock is higher, so it can move farther and therefore has a greater amount of this potential energy. Students can readily learn this idea, just as they learn about gravity, which will cause both rocks to gain speed as they fall.

Do remember, however, that it is less crucial that elementary school children memorize the definitions for matter and energy than it is that they understand *how to think* about making order from the diverse materials we encounter in nature.

Doing science sometimes means getting messy. Here some younger students explore the properties of soapy water.

Elizabeth Crews/The Image Works

A class where neatness rules

New teachers intimidated

"Signs of important activity"

"Oh, yes," the principal replied.

I knew I was in trouble at that school. My science projects tend to make the children excited and rather noisy. When I saw the classroom, my fears were confirmed. Not only was the place too quiet, it was too neat. I thought that meaningful exploration could not possibly be taking place if the room was dead quiet and perfectly neat.

Sometimes new teachers who want to engage their students in a student-centered, activity-based science learning environment are intimidated by messy materials and by what appears to be disarray. It is true: when students are using materials, working in groups, and trying to solve their own problems, the classroom will look a bit messy and sound a little noisy. But those are good signs. They are indications that active learning is taking place and that the students are taking charge of their own experiences—with guidance and coaching from their teacher.

Think back to the story in Chapter 4 about Mr. Wilson's students who were working in groups with icicles—weighing them, melting them, weighing them again, and so on—while Mr. Wilson walked around the room to make suggestions and ask questions. During this activity, do you think the room looked neat and perfectly arranged? I can tell you that it didn't. Some of the melted water spilled; shirt sleeves got wet; and for a while it seemed that pie pans and wadded-up paper towels were scattered everywhere. But these were signs of important activity.

Sometimes a messy room is part of doing science with children. So is a time for cleanup. When an activity is completed for the day, students can have assigned tasks that make the room cleanup easier to handle.

Potential criticism

Occasionally, if your room grows somewhat messy or noisy during a science activity, you may encounter criticism from an administrator or a parent. If so, be prepared to explain what the students are doing and why, and don't be intimidated. One kindergarten teacher in Vermont wanted to use sand and water at her class's science corner. She thought these items would help students learn some of the relationships involved in mixing and separating materials; also, seeing two materials that both pour but are quite different from one another would lead to some interesting questions. But the teacher was told that such materials were too messy for her school. The sand could get scattered, the water would spill—too much of a problem.

The critics silenced

But after attending an elementary school science institute as part of her inservice education, this teacher returned to her classroom determined to provide the children with materials for messing around. Luckily, she discovered a magic word to appease her critics. "As long as I call it science," she later reported, "everyone is happy with all the messing around my kids do!" (Goldhaber, 1994).

In fact, the science education reform movement does support and recommend the use of materials and activities to engage students in doing science, and it is understood that the process will not always be silent and neat. In most schools today, a classroom that is strictly quiet all day means that one-way communication is going on, from teacher to children. Most educators recognize that is not an effective means for true learning to occur.

Mess = science at work!

Messiness in the Categories

After reading about Ms. Hager's class, you may be wondering why she involved her students in an activity that did not result in a clear classification for the strange green matter they created from cornstarch and colored water. What is the point of having first graders investigate something that is a little like a liquid and a little like a solid?

Engagement → difficult questions

First, remember that teaching science involves getting students excited about new materials and the ideas they provoke. We want students to be engaged, eager to explore their own questions and create possible solutions, thus building their own meanings. But engaging students in their

A Poetic Comparison

By fifth grade we may ask students to express the science ideas illustrated in this chapter in many different ways. One imaginative way is through poetry, as in the following poem in two voices.

Solids and Liquids

I am a solid.	I am a liquid.
I have a definite shape.	I take the shape of my container.
I am rigid.	I can flow.
My molecules are close together.	My molecules are farther apart.
I can become a liquid.	I can become a solid.
As I become a liquid, I take in lots of heat—	As I become a solid, I lose lots of heat—
NOW I am a liquid.	NOW I am a solid.

For students, this type of poem is often effective in making science comparisons. You could expand on the poem by adding a third voice for "I am a gas." What might that sound like?

Daring to keep it complex

WE SAND AWAY AT THE INTERESTING EDGES OF SUBJECT MATTER UNTIL IT IS SO FREE FROM ITS NATURAL COMPLEXITIES, SO NEAT, THAT THERE IS NOT A CREVICE LEFT AS AN OPENING. ALL THAT IS LEFT IS TO HAND IT TO THEM, SCRUBBED AND SMOOTH, SO THAT THEY CAN VIEW IT AS OUTSIDERS.

—*Lisa Schneier*

own thinking often exposes the complexities of subject matter. Students may come up with questions that they cannot answer, or that we as their teachers cannot answer. Sometimes they may even raise questions that the most accomplished scientists in the world have not been able to answer.

"Keep it simple," we often hear as a standard phrase of advice. Do we dare, then, as teachers to keep it complex (Duckworth, 1991)?

I think we can dare—and we ought to. This is an exciting challenge for elementary school science teachers. We must not run away from complexities. We need to help children feel comfortable with the known and the unknown. We want them to understand that the process of scientific exploration does not always lead to a single right answer. If they become convinced that science is just a series of collected facts about the natural world—facts without ambiguities and uncertainties—they have missed the point, and our teaching has missed the mark. To sum it up neatly: in the real world, science sometimes offers answers that are *not* neat.

In the case of classification, what we know about children's development suggests that they gradually learn more complex classification systems as they mature (Siegler, 1991). How useful, then, to challenge them with green goo—a type of matter that can fit into two categories at the same time! This type of classifying activity can help prepare them for more sophisticated classifying schemes as they continue their science experiences in elementary school and beyond.

The green goo challenge

The tendency to oversimplify science curriculum does not serve the best interests of children as learners. It does not invite them to probe their ideas and examine their thinking. Instead, I recommend inviting your students into the brainstorming processes that allow them to construct new ideas for themselves—even if these processes are not always neat and predictable.

The next science story takes us to upstate New York, where I had the opportunity to cook scrambled eggs while eighteen first graders watched, guided, and participated. Certainly the principal from the midwestern elementary school would never have coded this activity as "scientific." See what you think.

SCIENCE STORY

States of Matter and Scrambled Eggs: An Eggsperiment

I am visiting a first-grade classroom in an affluent northern suburb in upstate New York. The teacher, Ms. Miller, has invited me to work with her first graders as they explore states of matter. It is near the end of the school year, and the children are completing their matter unit. They are bright and very eager learners.

When I enter the room, I find a work table, covered with a plastic cloth, set up for my visit. It has an electric frying pan, two eggs, a stick of butter, a plastic knife, plastic cups, a double-pan balance scale, and some plastic cube weights. Ms. Miller and I have discussed my plans beforehand, and we have obtained permission from the principal to use the electric frying pan in the classroom.

Familiar materials

This is not my first visit to Ms. Miller's class, so the children feel comfortable chatting with me. The first thing I ask them to do is to name some of the things that we will be working with. The children volunteer the names for all the items, and Ms. Miller lists them on poster paper under the heading "Materials for Our Experiment."

One girl blurts out, "Ms. Miller, that should be EGGsperiment instead of EXperiment." Ms. Miller changes the spelling, and we all laugh.

A corrected title

My thinking: I have decided on the egg-scrambling activity because I saw a kindergarten teacher demonstrate it for a graduate class that I teach. I love the idea of using materials from the kitchen in the classroom. These are materials the students should feel comfortable with, and they offer a clear connection to the students' daily lives.

I ask the children what they think I may be doing with these materials, and they all respond, "Making scrambled eggs." "Yes," I say, "and as we prepare the eggs, we are going to make careful observations of the butter and the eggs and how they change."

I crack the eggs in a plastic cup and stir them around with the knife. I ask the children what this looks like, and they observe that it is a yellow liquid. "How do you know that it is a liquid?" I ask. One student says, "You can pour it," and another adds, "It has no shape." Clearly they have been messing around with liquids before my visit.

Observing the eggs

"Can we weigh the eggs?" one student asks. "Sure," I reply, and we pour the eggs into a cuplike holder on our double-pan scale. Two students help put plastic cubes in the opposite pan, and the children slowly count, one at a time, until the scale pans balance at twenty-one cubes.

One student is curious.

I ask the class if they think the eggs will weigh the same after they are cooked. One student says they will weigh more because we are adding butter. Another student says they will weigh less because the cooked eggs are not as heavy. Many students say they don't know. Ms. Miller writes the following on the poster paper:

What is your prediction?

2 uncooked eggs weighed 21 plastic cubes

My thinking: I am pleased to have the students practice weighing objects and counting cubes. I am also interested in understanding what they think the outcome of a before-and-after weight test will be. Actually, though I'm supposed to be the expert here, I don't know the answer myself. As one student said, we are adding butter, but I also know that the eggs will lose much of their water content in the process of cooking.

Stumping the "eggspert"

As I hold the butter, I ask the children about its state of matter. While they seem to know it is a solid, one student named John says, "It has a liquid inside."

John's unusual idea

"Can you say more about that?" I ask.

"Well," he goes on, "when you leave it out, it gets soft because there is a liquid inside."

"What do you think about that?" I ask the class. The children agree that butter can be solid or liquid and urge me to put it in the frying pan. Before

Responding to John's idea

doing so, I cut it into several slices and ask the children to take notice of the inside of the butter. They cannot see a liquid *inside* the butter.

My thinking: I believe what John meant is that the butter is soft inside. When another student remarks that solids can be soft inside, John agrees, "Yes, that's what I mean." There is no way to know for sure, of course. In first grade, students do not always have the right words to express what they are thinking. They are still developing language skills for this type of communication. This activity will help them do that.

The students watch as the butter melts in the frying pan. "What state of matter is the butter in now?" I ask. The students all acknowledge that now it is a liquid. "That is what happens when a solid melts." I add, "It becomes a liquid." (I am adding my own language here.)

As the eggs begin to cook in the melted butter, I ask the children to observe carefully what seems to be rising from the eggs. Some children call it "gas," while others call it "steam." They are aware that this is an example of the state of matter called "gas."

"Where is it going?" I ask. The children say it is going up. "Yes, up into the air," I remark.

The children watch carefully.

My thinking: I am hoping to help the children realize that examples of the states of matter are all around us. They are also observing melting and convection— ideas I share with them without emphasizing the terminology. The effects of heat energy on the butter and on the liquid water in the eggs are not the point of this investigation. If the children want to explore the role of heat energy, we can do so, but the topic doesn't come up.

After the eggs are cooked, I ask the children what state of matter they are in now. Everyone says "solid." I ask why they think so. Several students say the cooked mass of eggs does not pour and does not lose its shape. One student says it is a soft solid, like butter. Then the students remind me that we have to weigh the cooked eggs.

Asking the students for explanations

We place them in the cuplike pan on the balance scale to weigh them. Once again, two children come up to place the plastic measuring cubes on the other side of the balance scale. The cooked eggs weigh 21 plastic cubes, the same as the uncooked eggs. Some children write that down in their science journals, and Ms. Miller writes it on the poster paper.

Outcome of the weight experiment

My thinking: I am surprised that the eggs weigh the same before and after cooking. I reason that the mass lost to steam has been made up for by the mass of the melted butter. That is a concept for a later lesson, however. Messy science, I think—the outcomes aren't always easy to explain.

Looking toward a later lesson

Just at that moment, the physical education teacher comes to the classroom to take the children to their gymnasium activity. The children start to moan. "Can we stay here?" they ask Ms. Miller. They want to draw and write in their science journals.

Ms. Miller assures them that they can record their observations after gym time. Later, she carries through with her promise, as shown by the journal reproductions in Figure 7.2.

FIGURE 7.2   The "eggsperiment": science journal drawings and notes from the students in Ms. Miller's class.   (a) Weighing the eggs on a double-pan scale. (b) One student's rendering of the three states of matter: liquid eggs in the cup, gas rising from the eggs as they cook, solid cooked eggs in the scale. (c) Another student's drawing of the eggs in the frying pan. (d) Anthony's inferences.

EXPANDING   MEANINGS
....

The Teaching Ideas Behind This Story

■ When I decided to scramble eggs with the children, I wanted them to examine the eggs from a particular perspective. They had another idea in mind: weighing the eggs. I followed that suggestion as a way to honor their curiosity and encourage them in their own thinking.

■ Even though I didn't know whether the eggs would gain or lose weight in cooking, it was an interesting question, so I let the children explore it. I learned something myself from the outcome. If science were always neat, results would always be predictable. But that is not the case, in either real science or in meaningful science activities with children.

■ The way the activity ended—with the children so eager to write in their journals—reminded me that when students are really engaged in their own thinking and their ideas are valued, they often do not want to move on to something else.

The Science Ideas Behind This Story

■ The phenomenon of cooking the eggs is more complicated than the melting of icicles described in Chapter 4. When the eggs begin to cook, the part of the eggs that is composed of water begins to evaporate as steam. Hence, the mass of the remaining egg is changed.

■ The hot steam is pushed up into the air through convection.

■ On the other hand, mass is gained when the melted butter combines with the eggs.

■ Therefore, it was a surprise—and something of an accident—that the mass turned out to be the same before and after cooking.

Questions for Further Exploration

■ What other types of cooking activities could demonstrate three states of matter?

■ Can you think of a follow-up activity for Ms. Miller's class to explore the idea that a liquid weighs less when some of it changes into steam?

Resources for Further Exploration

Electronic Resources

Kids and Family Channel. **http://www.infoseek.com/Topic?tid=12663.** Follow the links to "Kids' Activities" and "Science" for ideas about science explorations that you can do with children. (Tip: If the exact web address has changed, begin at the main Infoseek site.)

The Last Word. **http://www.newscientist.com/lastword/lastword.html.** A web site from *New Scientist* magazine with hundreds of questions submitted by readers—imaginative questions like those a child would ask, from "Why is the sky blue?" to "Why don't animals need glasses?" The answers, provided by many scientists, help demonstrate that science is not neat.

Physics Resources for K–12 Teachers and Students. **http://www.physics. ohio-state.edu/open_house/education.html.** This Ohio State University site has links to many web resources with interesting lesson ideas in the physical sciences.

Print Resources

Strongin, K., & Strongin, G. (1991). *Science on a Shoestring.* 2nd ed. Menlo Park, CA: Addison-Wesley. This book contains over sixty investigations designed for grades K–6. They are grouped under three themes: matter, change, and energy. Several of the investigations address how matter changes.

SCIENCE EXPERIENCES AS LIFE EXPERIENCES

As you can see from the stories and ideas in this chapter, science experiences can be complex, messy, and at times funny. Even so—and sometimes *because* of their messiness—science experiences can provide an exciting way to engage your students in thinking about the ordinary materials of their daily lives.

Implications for adult life

What they learn through "messy" science can have wide implications. After all, our lives are not always neat, simple, and absolute, and the world we live in is complicated. If we can help students become genuine explorers and investigators, willing to confront complexity and try out their own ideas for making sense of it, we will have taken a big step in preparing them for adult life.

As we proceed to the next chapter, notice how older students begin to carry their science investigations further, delving more deeply into the materials and experiences their teachers provide. Look for ways in which older elementary students can use their classroom investigations as springboards to further research in trade books, conventional textbooks, or on the Internet. As you do so, keep thinking about ways in which science connects with, and reflects, our daily lives.

Key Terms

classifying *(p. 130)*
property words *(p. 131)*
states of matter *(p. 139)*
solid *(p. 139)*
liquid *(p. 139)*
gas *(p. 139)*
suspension *(p. 140)*
physical science *(p. 142)*
matter *(p. 142)*
energy *(p. 142)*
element *(p. 142)*
atom *(p. 142)*
compound *(p. 142)*
molecule *(p. 142)*
mixture *(p. 142)*
kinetic energy *(p. 142)*
potential energy *(p. 142)*

8 Sustained Inquiry: Explorations of Living Things

FOCUSING QUESTIONS

- What experiences, if any, have you had with planting seeds, indoors or outside?

- What do you think it means to be "alive"?

- What is the point of a class science project that takes days or weeks to complete?

- How does working with a partner or small group affect your learning experience?

You may remember the first time you planted a seed and watched it develop into a plant with a stem and leaves and sometimes even a flower, depending on the type of seed it was. You may still be planting seeds in a home garden or window box. Plants and their growth and development remain a fascination for many home gardeners. All living things, including we humans, share the common thread of growth and development over time. This change over time is particularly exciting for students to explore.

In this chapter, we look at young children as they experiment with seeds and small living organisms. Then we examine how the activities can change as students reach higher grades. Notice from the stories how the complex relationships between living things and their environment are explored more and more deeply as students mature.

Especially in the stories of children in third grade and higher, you will see an emphasis on **sustained inquiry,** that is, the development of an extended investigation over a period of time. You'll see students working in pairs or small groups, collaborating with their teachers and with each other to exchange ideas and discoveries. Watch for evidence of three of the encompassing themes from the *National Science Education Standards* that we discussed in Chapter 1:

1. We learn science by inquiry.
2. We learn science by collaboration.
3. We learn science over time.

We begin with some brief stories about experiences I have had in discussing with young children the basic science question: "What does it mean to be alive?" These stories will set the stage for the more extended experiments described later in the chapter.

SCIENCE STORY

What Does It Mean to Be Alive?

The science idea of "living" versus "nonliving" has its own special magic for children, and they often wonder on their own what it is that makes something alive. There are many ways you can draw on this natural curiosity to spark a class discussion. In the following two stories, we'll see how some first and second graders approach the subject.

What Makes a Rabbit Real?

One day, I bring a velveteen rabbit to a class of first graders, along with the famous children's classic: *The Velveteen Rabbit* by Margery Williams. In this book, a stuffed rabbit comes to life in the eyes of a little boy who loves him so much that the rabbit is real to him. When the rabbit is old, a fairy makes him a *real* rabbit—"real to everyone." Released in the woods, the velveteen rabbit is a stuffed toy no more; he is at home with the other rabbits.

A classic story

I read the book to the class. They are absorbed and attentive—it really is a good story. Afterward, I hold up my own stuffed, velveteen rabbit with both hands. It has a button nose and no hind legs.

"In what ways is this rabbit like a real rabbit?" I ask.

Many of the first graders respond. "It is soft and brown like a real rabbit." "It has two ears." "It has two eyes." "It has a white tail." "It is furry."

"In what ways is it *different* from a real rabbit?" I ask.

Approaching a profound question

Again they have many good ideas: "It doesn't move." "It doesn't breathe." "It has no hind legs." "Its nose is a button." "It doesn't eat or drink water."

"What wonderful observations," I remark. "Let's think about what a real rabbit needs to stay alive."

My thinking: The question of what is alive is a profound one. I want to establish a major science idea here without "talking" it to the students, so I use same-and-different comparisons to help them make sense of the properties of the stuffed animal and a real animal. Besides exploring the science idea, I hope to build their process skills as they observe, reflect, and compare.

The students respond that a real rabbit would have to eat and drink and breathe. So they seem to know that food, water, and air are necessary for a real rabbit to stay alive.

"So what about this real rabbit?" I wonder out loud. "We know what it takes in. Does it give anything off?"

The students ponder this.

"Let's think about breathing," I say. "When we breathe in, does the air *stay* in?"

"No," one student calls out. "We breathe out too."

"And the rabbit?"

The students all agree that the rabbit breathes out just as they do. Two of the girls, who have petted real rabbits, remember that they could feel the rabbits' breath on their hands.

Then we discuss whether rabbits give off anything else. I'm looking for them to realize that if the rabbit takes in water and food, it must also give off wastes. One of the girls with prior rabbit experience suggests that a real rabbit would "go to the bathroom."

"Sure," I say, accepting this euphemism. "Why do you think it would have to?"

The children aren't sure. Then one student says that if a rabbit kept eating and eating without going to the bathroom, it would explode. Though others immediately label this idea as "gross," it seems that the idea that when something is alive, it gives off things to the environment and takes in things from the environment may be taking shape in their minds.

Pondering the characteristics of life

A "gross" inference

Are Plants Alive?

On another day, in a second-grade class, I have a similar discussion, but this time I've brought in a silk plant and a real plant. With plants, it is often more difficult than it is when exploring animals for young children to grasp what makes them "alive." (On the other hand, it is easier to have a real plant available than a real rabbit.)

I ask, "In what ways is the silk plant like the real plant? In what ways are they different?"

The children are quick to observe that we do not have to water the silk plant, and it cannot grow. It always remains the same size and color. Some children remark that the silk plant "won't die." That comment brings us to the idea that things that are not living cannot die.

Some children also say that the silk plant looks so real that it is hard to tell it is a fake. I ask, "What could you do to tell if it were living or nonliving?" The children respond that they could feel the leaves and then would know it was fake.

"How do real leaves feel?" I ask. Now the responses are limited. The children are not sure how real leaves feel, although one student says "waxy." In general, they just know this plant does not feel real.

They persist in their comparisons: "The plant is not real because we don't have to water it." One students offers, "It is not in real soil." Still another suggests that "nothing will happen to it in the dark"—getting back to the idea that it cannot die.

My thinking: The idea that living things grow and change develops with maturity. Many times the students hear the word *grow* used in the context of nonliving things, as in the phrase "growing crystals." I am hoping to nurture the idea that for something alive to grow, it needs resources—stuff from the environment around it. Hence, I am pleased that these students seem to get the idea that this silk plant requires no care—no "stuff."

Sometimes young children think that a real plant is not alive. "It is not moving," they may say. "It doesn't eat!" is another common response. This is an understandable alternative conception that develops as young children apply ideas about animal life to plants. In such cases, it is useful to remind the children that since the real plant needs materials from the environment—like water and sunlight—in order to grow bigger, the chances are that it is alive.

> **CHILDREN'S IDEAS ABOUT THE CHARACTERISTICS OF ORGANISMS DEVELOP FROM BASIC CONCEPTS OF LIVING AND NONLIVING.**
>
> *—National Science Education Standards*

How do we know if a plant is alive?

An alternative conception of life

EXPANDING MEANINGS

The Teaching Ideas Behind These Stories

■ A stuffed animal or artificial plant makes an excellent prop for discussing the nature of living things. However, it is also helpful to have a real counterpart available for the children's observations. The teaching dea here is to provide opportunities for observation and comparison.

■ Students come to school with many ideas about what is living and what is not living. Often, though, some of these ideas are alternative conceptions. By allowing students to express their ideas, you can discover these alternative conceptions and help the students to change them.

■ In addition to using props like stuffed animals, you may want to ask students to name and discuss living and nonliving things in their own environment.

The Science Ideas Behind These Stories

■ Living things are distinguished from nonliving things by their interactions with the environment. That is, they take in materials from the nonliving environment, and they give off materials to the environment.

■ Using the materials they take in from the environment, living things carry on life processes that give them the energy for growth and development.

■ Without these interactions with the environment, living things could not grow and change over time.

Questions for Further Exploration

■ If no rabbit is available in your school, what other animal(s) might you use?

■ What are the wastes given off by green plants?

Resources for Further Exploration

Electronic Resources

Living Things. **http://sln.fi.edu/tfi/units/life**. This web page, sponsored by the Franklin Institute in Philadelphia, has numerous links to information about plants and animals and simple investigations.

Plants (1997). Chicago: Clearvue. Designed for children in the elementary grades, this CD-ROM explores plants in their natural environments as well as the interactions of plants and animals. Available from CCV Software, Charleston, WV; **http://www.ccvsoftware.com**.

Print Resources

Brainard, A., & Wrubel, D. (1993). *Literature-Based Science Activities: An Integrated Approach.* New York: Scholastic Professional Books. As noted in Chapter 6, this book offers ideas for using children's literature in science activities.

Richter, B., & Nelson, P. (Eds.). (1994). *Every Teacher's Science Booklist: An Annotated Bibliography of Science Literature for Children.* New York: Scholastic Professional Books. This book contains excellent listings for literature connections as well as science book titles in all areas of elementary science, including the science of living things.

PLANTS AND ANIMALS IN YOUR SCIENCE CORNER

Often, as with the velveteen rabbit and the silk plant, we can bring objects or animals from outside the classroom to use in activities concerning the properties of living things. But it is also important to have living plants and animals in your classroom on a daily basis—as many as you are comfortable having and as local regulations allow.

Practical animals for the classroom

Some localities have restrictions based on concerns about allergies or infections, not to mention the difficulties posed in caring for classroom pets. Even so, there are usually several options. For instance, you can keep fish of many varieties; children frequently marvel at these vertebrate creatures that live so comfortably below the surface of the water. You can also have various types of invertebrates in your classroom. As you will discover in a story about snails later in this chapter, invertebrates tend to be easier to care for and are subject to fewer restrictions than vertebrates.

Having plants and animals in your science corner makes them part of the daily life of your classroom, so the children will be observing and caring for them regularly. These examples of living things can also help connect science to your students' environment outside the school, especially if the students themselves are encouraged to bring in plant materials that interest them. (It is probably not a good idea to encourage them to bring in animals!)

Connecting to the students' environment

Finally, keeping living things in the science corner can build the children's enthusiasm and provide opportunities for more extended activities, such as those described in the following stories.

SCIENCE STORY

From Seed to Plant: A Failed Experiment

Over the past week, Mr. Bauer, a first-grade teacher in a city environment in the Midwest, has kept some popcorn kernels in the class's science corner, with a card asking students to identify them and think about where they come from. Now he collects a wider assortment of edible seeds—pumpkin seeds, sunflower seeds, and pea seeds—for his students to explore and plant. He distributes cups of the seed mixture, and the children work in pairs to explore these objects.

"What are all the things you can say about these things?" Mr. Bauer begins. The children respond mostly with property words: "Some are green." "They are small." "They are hard." "You can eat them." "You can cook them." "Some are white. Some are yellow."

Observing properties of seeds

One student, Bethanne, says, "They're seeds. You can plant them, and they will make a new plant."

The students agree that seeds can grow new plants, though some are not sure that popcorn and peas belong in this category.

Mr. Bauer asks, "Where do you find these things?"

Some of the children answer, "The supermarket," and Mr. Bauer then probes, "Well, where does the supermarket get them?" The children aren't sure. Some say that corn seeds come from corn plants and that pumpkin seeds come from pumpkins, but they are unsure how the seeds got to the supermarket.

Mystery of the supermarket

Students marvel when a small seed germinates and produces a new plant. They compare their seed growth with each other as they explore life cycles.

Richard Hutchings/Photo Edit

Mr. Bauer asks the students to draw pictures of their seeds. Some label the drawings with the names of the seeds.

Mr. Bauer's thinking: Mr. Bauer is interested in having the students explore the seeds before they select one to plant in a cup of soil. He wants to encourage the children to notice things about seeds and to reflect on where they come from.

Mr. Bauer explains, "We are going to plant our seeds. What will we need?" The children say that they will need a pot of soil, plus water. Some say that they will need to place the pot in the sun. Their prior knowledge definitely includes some basic understanding of plants.

Mr. Bauer now invites each student to take one seed apiece and plant it in a paper cup filled with soil. The children do so. With Mr. Bauer's guidance, they place their seeds about 2 centimeters below the surface, moisten the soil with water, and place the cups on the window ledge of the classroom. In the following days, they check the cups to see what is happening.

Planting the seeds

The experiment meets with great success. All the cups sprout a new plant—except for one cup that belongs to a student named James.

Success!—with one exception

Mr. Bauer's thinking: Worried that all the seeds would not germinate, Mr. Bauer has planned for this eventuality by planting several cups at home. Early one morning, he makes a quick substitution, replacing James's empty cup with a cup that has a little seedling in it. He wants James to experience the same success as the other students.

The teacher's concern

That day, James appears amazed to find that his seed has finally grown. Mr. Bauer is delighted at James's reaction, but also puzzled at James's absolute astonishment. Nevertheless, in the afternoon, Mr. Bauer remarks to James that seed growth is often seen as a great miracle and he can understand James's joy.

James's astonishment

"It really *is* a miracle," replies James. "I ate the seed!"

SCIENCE STORY

What's Inside a Seed?

I am visiting Ms. Fraser's third-grade class in a suburban area in the Southeast. I have been working with Ms. Fraser and other teachers in the school as they implement a series of science lessons they have developed.

Ms. Fraser's third graders have been exploring plants that grow in their community. Their science corner is full of examples. Today they are going to examine a seed of one such plant, inside and out.

Ms. Fraser has soaked some lima beans (called "butter beans" in this southern community) overnight. She has provided paper towels, foil pie pans, and magnifying lenses for students to use. She also has a batch of beans that she has *not* soaked overnight. She holds these up to the class, and all the students recognize them as butter beans.

"Why do you suppose these butter beans are hard on the outside?" she asks.

Students observe and infer.

One boy answers, "To protect the baby plant inside."

"How do we know there is a baby plant inside?" Ms. Fraser asks.

"Well, it's a seed," the same boy replies. "Seeds have baby plants inside."

Ms. Fraser challenges him, "Okay, but we are doing science, and we need proof. What shall we do to prove your idea?" At this point a girl says, "Some seeds don't have a baby plant inside. They're empty."

The challenge

My thinking: Good point, I think. This reminds me of Mr. Bauer's problem when he worried that some seeds would not sprout.

Ms. Fraser asks the students how they can find a baby plant inside the bean, and they understand that they need to open the bean up. Now she shows the class the soaked batch of butter beans, and they notice how different these look from the unsoaked beans. (When a lima bean is soaked in water, the seed coat, a thin protective outer layer, peels away. The lima bean then is softer and obviously has two halves.)

Ms. Fraser divides the class into pairs of students and invites each pair to take two or three soaked beans, some paper towels, a pie pan, and a magnifying lens. Ms. Fraser asks them to draw pictures of the insides of their beans. Also, pointing out some books about seeds that she has gathered from the class library and the school library, she invites the students to label their drawings.

Ms. Fraser's thinking: Ms. Fraser wants the students to look up the names of the parts of their seeds. For this purpose, she has provided resource books to use. To encourage the children's own research, she does not tell them which parts to label or where exactly in the books to find the names they need.

As each pair of students examines the inside of a lima bean, they notice a tiny leaflike structure: the baby plant itself. Their magnifying lenses help them see the details of this baby plant. They draw it and label it "embryo," as it is called in the reference books. It is embedded in the two thick halves of the bean seed. They label these thick halves **cotyledons,** again following the books (see Figure 8.1). The science books have large, clear drawings of the inside of a lima bean seed, such as the one on the following page. They look for the drawings to find the labels.

Ms. Fraser visits the students as they are working. "What do you notice?" she asks. "Why do you suppose there are two halves to the seed?" She asks questions and prompts the students as she visits.

"What do you wonder about?" she asks. One pair of students wants to know how the seed can produce a new plant. Ms. Fraser addresses this question to the class, explaining that they can have the opportunity to observe how the seed grows into a new plant right inside a plastic bag. The students are intrigued and excited.

The Following Days: Germination Bags

The next day, Ms. Fraser brings another batch of presoaked butter beans to class. She invites the students to join their partners and take a couple of presoaked beans, a small Ziploc plastic bag, and some paper toweling to their desks. The following instructions, written on a chart that is placed at the front of the classroom for all the students to see, guide them:

> Mixing hands-on work with book research

> The baby plant revealed

> Ms. Fraser's questioning

FIGURE 8.1
....

The parts of a seed, revealed by dissecting a soaked lima bean.

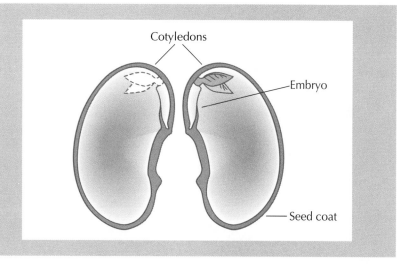

Cotyledons

Embryo

Seed coat

Place a flat, moist paper towel in the bag, folded, like a lining for the bag. You may get some water from the sink in the classroom. Make sure the towel is not dripping wet. Place two presoaked butter beans inside the bag between the paper towel and one side of the bag. Put a staple on either side of the seed to hold it in place.

Instructions for germination bags

Ms. Fraser herself uses a large Ziploc bag. In this one bag she places a lima bean, a pumpkin seed, a bush bean, and a corn seed on the moist paper toweling, stapling the seeds in place so they will not shift in the bag (see Figure 8.2).

Ms. Fraser's thinking: While the students are setting up their investigations with their lima beans, Ms. Fraser will model the germination of several seeds, so the students see that seed growth has certain common properties regardless of the type of seed.

The teacher's modeling

Ms. Fraser asks the students to decide where they will hang their germination bags. **Germination,** she tells them, is what we call the process when seeds sprout and begin to grow. Do they want their bags to hang in the well-lit classroom on the cork board—or in the dark, inside a closet? Should the plastic bag be laid down on a shelf? On the window sill? Somewhere else?

Choices for the students to make

My thinking: I notice that the only rule Ms. Fraser has given the students for the care of their bag and seed is that they should keep the paper towel moist. The other choices are for them to make, although she is helping them see the options.

FIGURE 8.2
....

**Ms. Fraser's
germination bag.**

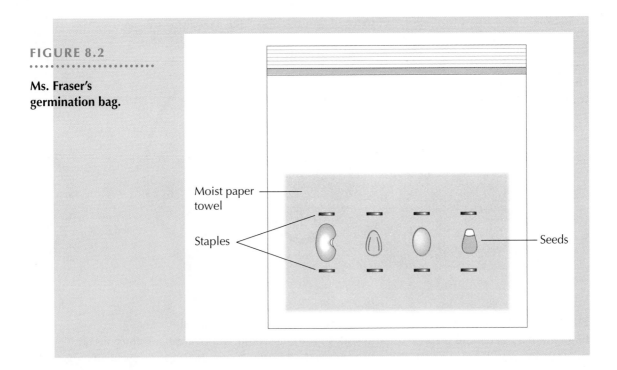

Moist paper towel

Staples

Seeds

After placing name labels on their bags, the pairs of students engage in extended discussions about where to hang the bags. Some students insist that the seeds need light, while others believe seeds will grow in the dark. One girl points out, "It is dark under the soil and the seeds grow there." Some of the partners disagree with each other, but they work out compromises. In the end, some of the bags are hung with push pins on the cork board in the classroom, and others are placed inside closet doors where there is no light. Ms. Fraser suggests that all the students write down, in their science journals, exactly what they did and what they think will happen.

The students debate.

Ms. Fraser's thinking: Ms. Fraser wants the students to gather their observations over the coming days and reach some understanding about seed germination on the basis of their experiences with the seeds in the bags. She wants the children to design their experiment any way they choose. She knows that the conditions that are needed for a seed to begin sprouting are often subject to alternative conceptions—the most common of them being that seeds need light to grow. Privately, she hangs her own germination bag in a dark closet, just in case no one else chooses to place a bag in the dark.

RIDDLE: I AM VERY LITTLE. I CAN BE DIFFERENT COLORS. I GROW TALL. I MAKE PLANTS. WHAT AM I?
***ANSWER:* A SEED.**

—*Andrea P., third grade*

Each day the students observe evidence of germination. Like a miracle, the young sprouts appear, right inside the plastic bags. The roots grow downward, regardless of the position of the plastic bag. The young stem grows in the opposite direction from the roots. The students observe how thin and willowy the roots are as opposed to the thicker and stronger stem. The students come to learn that the paper towel should be kept moist, but the lighting conditions are unimportant.

Sprouts growing in a bag

The students continue to experiment with the amount of sunlight the baby plants need. In the following days, the plastic seed bags continue to hang all around the classroom. Each day the students monitor the seed growth. Each day the process of germination and growth is visible to them, no longer a well-kept mystery in the dark recesses of a pot of soil.

When the green leaves have emerged and the sprouting seeds can no longer be contained in the plastic bags, the students transfer each germinated seed to a cup of soil. They carefully place the roots in the soil and pack the baby plant in with their fingertips. They water their plants and observe the continuing growth.

After about ten days, some differences are evident. Although the seeds germinated and the early seedlings grew regardless of the light conditions, the plants with leaves do well only in the light. The students begin to observe that their new green plants will *not* survive in a dark place. They do not understand why as yet.

Inferences about plants and sunlight

On day 8 of this seed-growth investigation, a student poses the question, "Why will the seedling grow in the dark, but the baby plant with sprouted leaves needs sunlight?" This becomes a research question for the class, and the children look up answers in their resource books about plants and seeds. They find out the following information:

The class's research findings

- The young plant, or seedling, gets its nutrition from the cotyledons and does not yet need sunlight.

- When the cotyledons are no longer needed, they emerge as seed leaves attached to the stem.

- When the plant begins to develop its first green leaf, it should be placed in a cup or small pot of soil to anchor the roots. At this time the green leaf should be exposed to the sun, because now the plant needs to make food. Green plants use sun, air, and water to make their own food.

My thinking: At this point I believe the students still have a long way to go before they have a deep understanding of the science idea that green plants make their own food. What does it mean for a plant to make its own food? What does the food look like? Is it a steak? Potatoes? Soon Ms. Fraser begins to address such questions.

Taxonomy

By exploring **seed plants,** which scientists call **angiosperms,** Ms. Fraser's students are learning about one major category of plants. Angiosperms produce flowers that form fruits with seeds. Garden and wild flowers, weeds, grasses, grains, plants that produce vegetables, and shrubs and trees that lose their leaves in autumn are all seed plants.

But why, you may wonder, do we bother to label plants with names like *angiosperm?* The answer is that classification schemes for living things help us understand and study the life on our planet. These schemes, which are human inventions, allow us to sort, group, and rank the types of life that share the earth with us (Margulis & Schwartz, 1982).

An entire classification system is called a **taxonomy.** The noted biologist and science writer Stephen Jay Gould (1981) has remarked that "taxonomies are reflections of human thought; they express our most fundamental concepts about the objects of our universe. Each taxonomy is a theory about the creature that it classifies."

In other words, by creating a taxonomy in which pumpkins and lima beans fall into the same category called angiosperms, we are expressing ideas about those plants' relationships to one another and to other plants as well. Typically, the relationships expressed by today's taxonomies are based on ideas about evolution—how and when particular living things developed their current form.

No taxonomy is absolute. Consider the most basic grouping, the kingdom. In science, a **kingdom** is the largest, most inclusive level of classification of living things. When I was growing up, the living things we learned about were all placed into two kingdoms: the Animal Kingdom and the Plant Kingdom. Now, however, scientists classify living things into *five* kingdoms. The number changed because scientists have found out much more about what is alive than they knew a few decades ago.

The five kingdoms recognized today are:

Monera (monerans). Single-celled organisms including all types of bacteria.

Protista (protists). Single-celled organisms including algae, amoebas, paramecia, and others.

Fungi. Multicelled organisms such as slime molds, yeasts, and mushrooms.

Plantae (plants). Multicelled organisms that make all their own food through photosynthesis. Includes mosses, horsetails, ferns, trees, and all seed-bearing plants.

Animalia (animals). Multicelled organisms including plant eaters, meat eaters, and parasites. Animals range from sponges, jellyfish, and snails to fish, amphibians, reptiles, birds, and mammals.

Ms. Fraser introduces the term **photosynthesis** to the group. This is the process, she explains, by which green plants use carbon dioxide from the air and water from the soil, in the presence of sunlight, to manufacture molecules of glucose, a simple sugar, which is then used by the cells of the plant to make energy for the plant to carry on its functions, including growth. This chemical process, combining materials from the air and soil to produce new ones, takes place at a special site in the cells of the leaf called a *chloroplast*. The chloroplast contains the substance **chlorophyll,** which helps the process of photosynthesis to occur. Chlorophyll also gives the leaf its green color.

> Introducing scientific terms

When the plants make their own food, Ms. Fraser continues, they give off extra water vapor and oxygen to their surroundings. This is something we can see, she tells the class. "How can we see it?" many of them want to know.

My thinking: Photosynthesis may be difficult for these students to understand, but by suggesting an observation the students themselves can make, Ms. Fraser helps to tie the idea to their own experience.

Following her guidance, the children place a small green plant (somewhat larger and more robust than their brand-new plants) in a large glass jar. They cover the jar with its lid. After a day or two, they see water vapor droplets appear on the inside of the jar.

> Making a complex process visible

My thinking: I'm pleased that Ms. Fraser is introducing a very complex process at this early grade level. The students will now have some prior experience with this idea on which to build as they mature.

E X P A N D I N G   M E A N I N G S

The Teaching Ideas Behind These Stories

- In the story of Mr. Bauer's class, we see how simple experiments with seeds can engage students in the mystery of growth and development.

- We also see that edible seeds can represent a challenge for young children. The fact that many foods come from the ground and start as part of a plant is not taken-for-granted knowledge.

- The activity through which Mr. Bauer guides his students has many fine points. It also shows us, however, that even with the best intentions, it is not a good idea to fool students. Instead of trying to sneak in a replacement for a seed that didn't grow, it would have been better to avoid setting up an experience in which students might personalize the success or failure as their own success or failure. Mr. Bauer would have been better advised to use larger cups with two or three seeds. He was lucky that all the other children's seeds germinated.

- Science is about experimenting. Sometimes it works, and sometimes it does not work. There is often much to learn from failed experiments. They provide opportunities to change direction and explore other possibilities. For instance, if some seeds fail to grow, you can ask, "Why do some seeds germinate, while others do not?"

- Notice how Ms. Fraser gives her students multiple opportunities for observation. For example, she demonstrates the importance of soaking the beans by asking the students to observe what has happened to the beans soaked overnight. Hanging the germination bags in the classroom also allows the students many chances to make direct observations.

- Notice, too, how Ms. Fraser's third graders go beyond the examination of the seeds and move toward collaborating in planning their own investigations. She has structured the activity so that they can discuss with their partners, and with other students in the class, the best way to germinate a seed. Each pair can make its own plans, carry them out, and determine the results.

- As her students develop questions that cannot be answered through direct observation and experimentation at this stage, notice how Ms. Fraser directs them to resources—in this case, books in the classroom. Some elementary schools use science textbooks, which can also serve as resources.

- The entire activity in Ms. Fraser's class, with its multiple spin-offs, takes place over two weeks or more—a significant allotment of time. Later in this chapter, we will see more examples of how science learning requires sustained inquiry over time.

- Even if a science idea is very complex, a demonstration or activity that presents part of it can be useful in elementary school science. By showing her students that a small green plant can release water vapor inside a closed glass jar, Ms. Fraser is leading them toward the beginnings of understanding, and she is helping to make an abstract concept more concrete.

The Science Ideas Behind These Stories

- Seeds contain the baby plant or embryo and the food that the embryo needs to begin germinating. This food is contained within the part of the seed called a cotyledon.

- Some seeds do not develop properly, and some do not even have an embryo. Thus, it is always a good idea to plant more than one seed.

- Seeds do not need sunlight when they begin to grow. At first, they live off the stored food in their cotyledons. Once the new plant grows leaves, it needs sunlight to make its own food.

- When seeds sprout, the cotyledons become the first leaves of the baby plant, and they fall off as soon as the plant grows its first adult leaves. Some plants produce seeds with one cotyledon; these plants are called **monocots.** Plants that produce seeds with two cotyledons are called **dicots.** Grasses and grains are monocots. Flowers, flowering trees, beans, and vegetable-producing plants are usually dicots.

- For many seeds, soaking will begin the germination process. Compared to seeds, sprouts (young shoots) have a better chance for survival when they are planted in the ground.

- The soil must be firm and porous to support the roots of a seed plant. The soil supplies green plants with an anchor for their roots and space in which to grow.

- Dissolved minerals in the soil make their way to a plant through the water intake of its root system.

- Although the minerals from the soil are important, green plants do not get *food* from the soil. They make their own food through the complex process of photosynthesis. This is a chemical process in which the leaf, aided by its chlorophyll, uses water from the soil and a gas in the air called carbon dioxide to make a simple sugar called glucose and release another gas, oxygen. The carbon dioxide enters the leaf through openings on its underside called *stomata*. The oxygen gets released into the air through these same stomata.

Questions for Further Exploration

- Try planting citrus seeds like grapefruit or orange seeds. How well do you think they will grow in your own home? Do they surprise you?

- Buy wheat berries (typically found in health food stores). Fill a shallow pie pan with soil. Sprinkle the wheat berries on the soil and water well. See what happens.

- What do we mean by saying that green plants make their own food? How is this different from what animals do?

- Since green plants make their own food, what is the plant "food" sold in stores?

- How can bean plants be a model of a life cycle?

Resources for Further Exploration

Electronic Resources

Brenckmann, F. *Seeds of Life: The Wonderful World of Seeds.* **http://versicolores. ca/seedsoflife/ehome.html**. An award-winning web site full of fascinating information and pictures.

Photosynthesis. (1997). Chatsworth, CA: AIMS Multimedia. This CD-ROM examines various aspects of photosynthesis in everyday life. Available from CCV Software, Charleston, WV; **http://www.ccvsoftware.com**.

Print Resources

Kraus, R. (1945). *The Carrot Seed.* New York: Harper. This is a classic children's story, fine for grades K–2, about a tiny carrot seed and the plant it grows into.

Margulis, L., & Schwartz, K. (1982). *Five Kingdoms: An Illustrated Guide to the Phyla of Life on Earth.* New York: W. H. Freeman. A detailed discussion of each of the five kingdoms of living things and a fine resource for teachers.

National Sciences Resources Center. (1991). *Plant Growth and Development.* Designed for grade 3, these experiments are useful for students to learn how to plan and carry out controlled investigations. Available from Carolina Biological Supply Co., Burlington, NC.

Planting in a Vacant Lot

In one inner-city environment, many of the students walk past an empty lot strewn with garbage on their way to school. Ms. Monteiro, a fifth-grade teacher, notices this, and she thinks about the possibility of planting seeds and nurturing plant growth as both a class activity and a means of community involvement. Recently she has read the book *Seedfolks* by Paul Fleischman, a delightful story of a community of diverse immigrants in Cleveland who contributed their time and energy to convert a vacant lot into a garden.

Inspired by this book, Ms. Monteiro and her class write letters to the local newspaper and to city officials in order to get the trash removed. It takes several months, but eventually the trash is indeed carted away. Now the students visit the lot and claim it for their own.

With the permission of the city, this fifth-grade class turns the vacant lot into a garden. Topsoil is donated by a garden supply store in a neighboring suburb, and the students get to work making the ground ready. They have already noticed how dandelions and crabgrasses grew in this lot despite the trash, so they have high hopes for their own plants. The students do a lot of work preparing the soil and researching the conditions under which their seeds will germinate.

A transformation

The students have many decisions to make. Some want to plant vegetables; some want to plant flowers. Whenever possible, Ms. Monteiro allows them to make their own choices and select their own seeds. The final choices include lima beans, sunflowers, pumpkins, lentils, marigolds, and snapdragons. By early spring, the students begin sprouting the seeds in containers in the classroom. When the seeds have germinated and have well-formed roots, the students plant them in the ground.

Making their own decisions

Meanwhile, in their classroom, the students have examined the seeds and the leaves of their sprouting plants. They have studied germination and the role of the soil, water, and sun. They learn that outdoor plants have their own particular growing seasons but that they can simulate some outdoor conditions in the classroom.

Some students who have immigrated from the tropical climates of islands in the Caribbean are disappointed to learn that the harsh winters of their new home will not allow some of their favorite foods to grow. One boy orig-

Learning the effects of climate

FIGURE 8.3 **Steps in planting an avocado pit.** (1) After washing the avocado pit and peeling off the skin, place toothpicks in the seed and suspend it in a jar of water, pointed side up. Making sure the bottom of the seed is covered with water, place the jar in a dark place. (2) Wait until roots begin to grow and the stem is about 15 centimeters high. To thicken the roots, cut the stem about halfway up and return the plant to the dark for another few weeks. (3) Finally, when the stem has again grown to about 15 centimeters, place the seedling in a deep pot of soil so that the roots are completely covered and the top 2 centimeters of the seed show above the soil. Water the plant well.

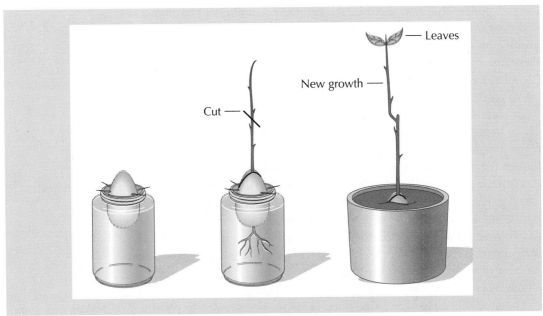

inally wanted an avocado tree in the garden. Instead, he plants an avocado seed in a carefully controlled environment in the classroom (see Figure 8.3).

As the season progresses, the students make regular visits to their outdoor garden. In their eyes, nothing has ever been more beautiful than this formerly vacant lot. A sign identifying the class and the teacher stands proudly on the site.

The project is limited by the brevity of spring in this midwestern city. But before the end of the school year, arrangements are made for the garden to be cared for during the summer by students and community volunteers. In the autumn, when the students return, they plant crocus and tulip bulbs donated by local businesses.

The end of the school year—but not the project

When Is a Vegetable a Fruit?

Ms. Byrne's third-grade class is learning about parts of plants that we can eat. One day she brings in a number of such edible plant parts: apples, oranges, celery, pea pods, lettuce, spinach, tomatoes, red and green peppers, mango, papaya, collard greens, broccoli, carrots, avocado, persimmon, and fennel. (Ms. Byrne knows that her students come from several different cultures and that several students have arrived from Central America, so she is sure to bring foods familiar to these children.)

After having the students divide into groups, Ms. Byrne invites them to use plastic knives to examine the specimens. "We don't know the names of all these fruits and vegetables," declares one student. "Yeah, I've never seen this before," another chimes in. Ms. Byrne explains (again echoing Richard Feynman's story in Chapter 3) that knowing the names wouldn't necessarily mean very much. "If all I knew about you," she says to one student, "was your name, then I wouldn't know very much, would I?"

To guide the students, she offers a suggestion: "We can begin by exploring which of our plant parts have seeds and which do not." The students begin to squeal with delight as they dissect their foods. John shouts, "This one is my favorite," as he holds up an avocado.

"What's that?" inquires a classmate.

The students begin to share information about their favorite foods while creating two groups of plants: those with seeds and those without seeds. John holds up the pit of the avocado and explains how to plant the pit in a pot of soil in order to grow a new plant. Ms. Byrne makes a note to do this in class the next day.

The children observe the seeds in one pile of foods and the absence of seeds in another pile. Holding up an apple and an orange, Ms. Byrne asks the students what we call these parts of the plant.

"They are fruits," declares Jessica, and everyone agrees.

"Yes, scientists refer to the parts of a plant that hold seeds as the fruit of a plant."

The children are stunned to notice, however, that the tomatoes, red and green peppers, pea pods, avocado, and persimmon are all in the pile with the fruits. "If a tomato is a fruit," asks Ruby, "then what are vegetables?"

Ms. Byrne responds, "What do you think?" She holds up the carrot and asks the students to think about what part of the plant it is. Someone knows it is a root. Others realize that celery is a stem; that broccoli florets

Foods never seen before

Puzzle: If a tomato is a fruit . . .

are flowers; and that collard greens, lettuce, and spinach are all leaves of the plant.

The students then take turns talking about the fruits or vegetables that they eat at home. Some take pleasure in describing the joys of eating mango or papaya or persimmon. Others explain how fennel or collard greens are prepared in their homes. This discussion leads into where the plant is grown, how it is eaten, and what significance, if any, the fruit or vegetable has in particular cultures.

Sharing information from different cultures

EXPANDING MEANINGS

The Teaching Ideas Behind These Stories

- In the vacant lot story, we see how nature can present itself in the most unlikely places. Plants break through wherever they can. These unlikely places can become precious resources to the urban teacher with a keen eye.

- The vacant lot story also demonstrates one way to involve the children with the surrounding community and vice versa, a subject we will discuss in more detail in Chapter 16.

- In the fruits and vegetables story, Ms. Byrne can create meaningful science experiences for her students because she knows who they are, what they eat, and where they come from. The experiences she provides help the students develop their personal knowledge of science.

- The story of Ms. Byrne's class also reminds us of the gifts of diverse student populations and how excited students get when the class curriculum reflects their lived experiences.

- In both stories, the students' ability to construct meaning is strengthened by their interactions as they collaborate with one another. They share knowledge, opinions, and observations.

- In the vacant lot story, the activity takes place over a long period of time—most of the school year and on through the summer into the next school year. This sustained project allows students to develop a deeper, more personal knowledge of the science ideas behind plants.

The Science Ideas Behind These Stories

- Fruits and vegetables are the edible parts of plants.

- In everyday language, fruits are usually those plant parts that taste sweet, particularly those we get from trees, such as apples, pears, peaches, oranges, and bananas.

- In scientific classification, however, sweetness is not the important factor. A **vegetable** is a root, stem, leaf, or flower of a plant. The **fruit** is the part of the plant that is a container for the seeds.

- To be even more specific, a fruit is the ripened ovary of a flowering plant.

- Nuts, corn, wheat, tomatoes, peas, beans, cucumbers, pumpkins, and squash are really fruits.

Questions for Further Exploration

- What are the fruits and vegetables that you encounter in your daily life?

- Are there any edible plant parts that you have difficulty labeling as fruit or vegetable? How can you find out the correct classification?

- In Chapter 10, we will learn about some students who use fruits and vegetables to make an edible model. What do you think it will be a model of?

Resources for Further Exploration

Electronic Resources

- Kimball, R., & Donoghue, D. (1997). *Botanical Gardens*. Pleasantville, NY: Sunburst Software. Appropriate for fourth-grade and older students, this problem-solving CD allows students to interact with conditions of a simulated environment and observe how they affect plant growth. **http://www.sunburstdirect.com**.

- *Ontario Agri-Food Education*. **http://www.oafe.org**. For various food-related classroom activities, click on "Teacher Resources" and follow the links.

Print Resources

- Fleischman, P. (1997). *Seedfolks.* New York: HarperCollins. This book describes a project to create a community garden.

- National Sciences Resources Center. (1992). *Experiments with Plants.* Similar to the volume *Plant Growth and Development* cited earlier, but the experiments are designed for grades 5 and 6. Available from Carolina Biological Supply Co., Burlington, NC.

A CLASSROOM INVERTEBRATE

Earlier in this chapter, we noticed young students experimenting with seeds. First graders planted them in cups, while third graders explored germination and some of the properties of green plants. Animal life is usually less accessible, and keeping an animal in the classroom often requires some kind of clearance from the school or the district.

Invertebrates vs. vertebrates

An exception to the constraints about using live animals with children is a small and lovely invertebrate animal called a land snail. **Invertebrate** animals have no internal backbone, and their skeleton is usually on the exterior of the body rather than the interior. Insects, jellyfish, worms, crayfish, and crabs are all invertebrates. **Vertebrate** animals, in contrast, are those with backbones, including all fish, amphibians, birds, reptiles, and mammals.

The fascinating, commonplace snail

The land snail has many interesting features to observe. Between spring and fall, in many temperate climates throughout the United States, large numbers of land snails can be seen in gardens and wooded areas. They are quite common. They have beautiful, patterned shells, and they secrete a silver trail as they move slowly in moist, shady environments. Land snails may be ordered from biological supply companies (see Appendix B). They reproduce easily because they are hermaphroditic; this means they have both male and female reproductive structures.

Making a home for your snail

Land snails are easily kept in the classroom. Here are some guidelines for preparing the snails' home:

1. Get a glass or plastic tank or terrarium. Alternatively, you can use a gallon jar or plastic container with a lid into which you have poked holes.

2. Cover the bottom of the tank or container with 2 centimeters of gravel.

3. Add soil to a depth of approximately 5 centimeters. Keep the soil moistened.

4. Add a rock, some moss, and some bark-covered twigs.

"I Am Living"

As we have seen in earlier chapters, students can write dialogue poems as a way to talk about their understanding of science ideas. The following poem in three voices, written by three teachers in a summer science workshop, sums up some ideas about the characteristics of living as opposed to nonliving things.

I am an *animal*.	I am a *plant*.	I am a *rock*.
I am living.	I am living.	I am nonliving.
I take in oxygen.	I take in carbon dioxide.	I do not breathe.
I need water to survive.	I need water to survive.	I don't need water to survive.
Feet and muscles help me move.	Wind helps me move.	Sometimes water helps me move.
I get my food from other living things.	I make my own food.	I don't need food.
With time, I grow bigger, smarter, and stronger.	With time, I grow bigger and stronger.	With time, I grow smaller.
A mate helped me reproduce.	A bee helps me reproduce.	I get no help! I don't reproduce.
I have a brain.	I don't have a brain.	What's a brain?
Someday I will die.	Someday I will die.	I could last forever.
I AM AN ANIMAL.	I AM A PLANT.	I AM A ROCK.

You may want to encourage your students to write similar poems. For a collection of dialogue poems focusing on insects, see *Joyful Noise: Poems for Two Voices* (New York: Harper and Row) (Fleischman, 1988).

Reprinted by permission.

At varying stages of development and learning, students will delight in exploring these tiny creatures. For younger students, land snails may be fascinating to keep in the science corner for observations. By fourth grade, students should be ready for the responsibility of handling these living organisms for more detailed study. At that stage, I like to have students work in pairs to share observations and inferences, each pair with its own snail. First, though, it is important that you explore the snails for yourself, with a colleague if possible, and record as many observations as you can about these creatures.

Snail explorations

To prepare the students to observe the behavior of their snails, distribute magnifying lenses and paper toweling beforehand. Once they begin to explore, you will probably hear squeals of delight fill the room. The land

What is alive? When students explore the properties of the garden land snail, they begin to gain an understanding of what it means for living things to exchange materials with their environment.

Frank Collins/
Annette K. Goodman

snails are adorable creatures. They are soft-bodied organisms that, when recoiled in their shells, can range from 3 to 5 centimeters in length, but when fully extended outside their shells can reach a length of 10 centimeters. The snails have four frontal antennae. Upon examination, you will learn that they have eyes at the ends of two of these antennae. The students get quite excited when they discover the snails' eyes.

You may want to ask your students, "How can we compare the size of our land snails with our own size?" By fourth grade, many students can compare the sizes of two objects by using ratios. As your students explore, they will also come up with many fascinating questions of their own, as the following story demonstrates.

SCIENCE STORY

A Book of Snails

Ms. Sigursky's fourth-grade class has 24 students and 14 land snails. Ever since the snails arrived and the students helped set up the terrarium, they have been eager to take the snails out.

Before they begin to handle the animals, however, Ms. Sigursky points

Observations of Snails

The left column below lists some observations of land snails that teachers made during an inservice workshop as they worked in pairs. The right column lists some observations made by fourth graders. How many similarities do you see? What are the key differences? On the basis of these comments, how do you think the teachers could mediate the learning experience for the fourth graders?

Teachers

"I love to watch the land snails. They really enjoy lettuce. When I place the lettuce near the snail, it stretches its antennae out. It eats very quickly. Their antennae look like little walking sticks."

"All of the snails' movements seem so well planned. They have extremely flexible bodies. I saw one of the snails bend his head to reach up to the top of the container. It has such strong suction. It pulled itself onto the bottom of the roof of the container and stayed there—upside down."

"The snails like to be upside down."

"When they move, they seem to secrete a clear fluid that they leave behind them. This must help them slide along the surfaces they move on."

"They follow hand movements."

"The snails sleep a lot. They move slowly and they don't make noise."

"When you touch the snails, they seem to shrink away."

"They are really graceful."

"They have different patterns on their shells."

"They are slimy."

"They excrete a green, slimy substance."

"I love to watch their bodies moving as they glide across their container. It almost looks as if there are waves inside of them."

Fourth Graders

"They're slimy."

"They leave a trail when they move."

"The green stuff is what they give off when they go to the bathroom."

"They like to hang upside down."

"They climb on each other."

"They're cute!"

"They like the light."

"They love lettuce."

out that working with live specimens is a big responsibility. She invites the students to brainstorm about how they will interact with the creatures. They come up with a number of rules to write on a piece of poster board, including the following:

Rules that the class brainstorms

- Do not squeeze the snails too hard.

- Handle the snails gently.

- Do not hold them up too high.

- Keep the snails well fed.

- Keep the soil in their cage wet.

- Keep the tank clean.

- Do not disturb the snails unnecessarily.

"Those are good rules," Ms. Sigursky tells them. "I think we're ready to begin our explorations." She asks the students to pair up with their science partners, explaining that each team will get one snail to work with.

"That leaves two extra snails," Maria calculates.

"Those are for me," Ms. Sigursky replies. "I forgot to bring lunch."

"Ewww!" several class members cry. Ms. Sigursky's bad jokes have become a staple of classroom humor. Nevertheless, Maria adds one more item to the list on the poster board: "Do not let Ms. S. eat snails."

One more rule

That day and the next, the students make their initial observations, writing notes in their science journals. Noticing the way the snails seem to move toward some things and away from others, they decide, with Ms. Sigursky's guidance, to conduct an investigation. The students carefully place their land snails on a flat surface, shine a flashlight in their direction, and watch the result. Then they place a piece of lettuce in the snails' path and observe the creatures' behavior. They do the same with drops of water. The students record these procedures and their observations, learning firsthand about the preferences of the land snails for the stimuli of light, food, and water.

What attracts a snail?

In the next few days, these fourth graders generate many other snail questions. These are a small sample:

The students' questions

How do you tell the sex of a snail?

What is the shell made out of?

How do they grow?

How long do they live?

How do they reproduce?

How do they breathe? Digest food?

How are land snails different from water snails?

What happens if their shell breaks?

Do they lay eggs, or do they have live baby snails?

How do they hang upside down?

Ms. Sigursky divides the students' questions into two groups: (1) questions they can answer through their own investigation and exploration, and (2) questions they can look up. (Which questions of those listed would you select for each category?)

The students remark that it will take a long time for them to answer some of their questions, and they wonder how long the class will have for snail observations. Ms. Sigursky responds that they can have as much time as they need.

The students turn to many resources to supplement their own observations: the Internet, resource books, CD-ROMs, friends who have snails as house pets. By the end of a month, the class begins to produce its own book about snails, full of careful drawings, observations, and information drawn from their own research.

A month-long investigation

After some debate about the best title for their book, the students decide to call it, *Snails Are Too Cool to Eat*.

E X P A N D I N G   M E A N I N G S

The Teaching Ideas Behind This Story

- Inviting students to examine living things in detail provides them with a fuller understanding of the question, "What is alive?"

- This sort of exploration can take many forms in the elementary school classroom. A wise teacher, like Ms. Sigursky in this story, creates an environment in which students' own ideas flourish.

- Exploring the snails in pairs encourages the students to collaborate and share ideas. This points again to one of the key themes from the National Science Education Standards: We learn science by collaboration. This is true because science is, among other things, a social process.

- The theme that science takes time emerges here as well, as it did in Ms. Fraser's and Ms. Monteiro's classes. Ms. Sigursky is generous in her promise of time for snail studies, confident that she can help the students make good use of the time they spend.

The Science Ideas Behind This Story

■ When you order land snails, biological supply companies provide complete instructions for their care and maintenance. Still, it is important to have the students themselves brainstorm about key aspects of snail care, as Ms. Sigursky's class did.

■ In addition to lettuce, land snails like oatmeal flakes with a little milk and a raw carrot or potato. Moistened bran flakes or bread may also be offered. It is important to sprinkle the oatmeal flakes with calcium carbonate once a week as a way to provide calcium to your snails.

■ Like many other invertebrates, snails are soft-bodied animals that have their hard parts on the outside as protection.

■ Other examples of animals with hard protective exteriors are crustaceans (such as lobsters, shrimp, and crabs) and arachnids (such as spiders and scorpions). The most numerous such animals, however—in fact, the most numerous of all invertebrates—are insects; they have penetrated all environments on earth. Insects, arachnids, and crustaceans are all called **arthropods.**

■ Snails belong to a group of invertebrate animals called **univalve mollusks.** Other univalve mollusks are the slug, conch, and abalone. The land snails and slugs are the only univalve mollusks to live on land.

■ The term *univalve* refers to the fact that snails have only one shell. Other mollusks, like clams, scallops, oysters, and mussels, have two shells and therefore belong to the group of invertebrates known as **bivalve mollusks.**

Questions for Further Exploration

■ Could you find a land snail in your environment?

■ What other small invertebrates could you find where you live?

■ What other interesting snail studies can you think of?

■ What are some of your own questions about snails?

Resources for Further Exploration

Electronic Resources

Goodman, A. K. *The Snail Pages.* **http://www.geocities.com/Heartland/Valley/6210.** Photos and information about land snails, including interesting links, from a collector of giant African land snails.

Liu, K. *Eye to Eye with Garden Snails.* **http://outcast.gene.com/ae/AE/AEC/AEF/1994/liu_snails.html**. A series of elementary school snail lessons with many good ideas and lots of background information.

Print Resources

Barrett, K. (1986). *Animals in Action.* Berkeley, CA: Lawrence Hall of Science. GEMS (Great Expectations in Mathematics and Science) inspired these activities designed for children in grade 6. They engage students in design and evaluation of their own animal behavior experiments, including experiments with land snails.

Buholzer, T. (1987). *Life of the Snail.* Minneapolis: Carolrhoda Books. This is a wonderful resource book for any young person interested in collecting information about land snails. The photographs are wonderful too.

Hickman, P. (1993). *Wetlands.* Toronto, Ontario: Kids Can Press. A publication of the Federation of Ontario Naturalists, these reading selections and projects include the study of snails as wetland wildlife. Designed for grades 3–6, the book also explores environmental considerations for wetlands.

Lionni, L. (1968). *The Biggest House in the World.* New York: Pantheon Books. A snail's father advises him to keep his house small and tells him what happened to a snail that grew a large and spectacular shell.

Ryder, J. (1982). *The Snail's Spell.* New York: F. Warne. The reader imagines how it feels to be a snail. A super story for all elementary grades.

WORKING TOGETHER TO CONDUCT INVESTIGATIONS OVER TIME

In this chapter, we have visited several classrooms and noticed different types of explorations with seeds, plants, and small animals. One way in which these stories converge is through the honoring of the students' own

Encouraging students
to pursue their
questions

Sustaining
investigations

How do you keep
students interested?

**AT ALL STAGES OF
INQUIRY, TEACHERS
GUIDE, FOCUS,
CHALLENGE, AND
ENCOURAGE STUDENT
LEARNING.**

—*National Science Education
Standards*

Students—and their
teacher—as
collaborators

ideas, a theme stressed throughout this book. In particular, the students in this chapter's stories received encouragement to research their own questions, even when that meant looking beyond the classroom on the Internet, in textbooks, in trade books, or down the block in a vacant lot. This is the basis of any true inquiry in science, and it is the way students can develop their own inquiry skills.

You also noticed, no doubt, that many of the experiences described in this chapter were time intensive; that is, the students were given plenty of time to explore their germinating seeds, their snails, their fruits and vegetables. Plants and animals change over time, and in order to develop students' understanding of these living things, we need to encourage the students to sustain their investigations over extended periods of time. In fact, you may often want to encourage your students to repeat experiments. This type of repetition is common to scientific research. Sometimes the second experience with the investigation helps students to notice things they may have missed the first time around.

Following through on science explorations over time is especially important for students in today's culture of "quick and easy." How, you may ask, can you maintain the students' interest in a single investigation that takes days, weeks, or longer? I believe you will find that when you ask the students to brainstorm the questions they have about their investigations, they become very interested in pursuing the answers. And as the stories in this chapter have illustrated, new questions often arise as the investigation continues. Sometimes, in fact, students want to research more questions than there will ever be time for in your class. As you get to know your students better with time, you can help them select the questions they are most "dying to answer."

Part of your role is to guide your students in their understanding of what can be explored through direct experimentation and what may require research from outside sources. You also take the responsibility to point things out and ask them to tell you what they have done so far. You guide and monitor and make suggestions. At the beginning of the school year, your students may need a good deal of structure, depending on how much experience they have had in working on their own. As the school year progresses and they get more practice in self-directed activities, they will become more able researchers.

The stories in this chapter have also stressed the collaborative nature of investigations. The students worked in pairs or groups, sharing their ideas. The teachers, too, essentially collaborated in the experiments, providing the topics and the materials but then allowing the students to gain control over their own learning.

Collaboration works best when the exploration is arranged so that each group of students can participate in the efforts and work of *all* their classmates. Often a "reporting out" time becomes the best part of student collaboration. As we will see in Part Three, it is also a wonderful way to assess what students have learned.

As students mature, they are able to plan their own investigations in more detail and manipulate experimental conditions to gather new information. In the next chapter, we will see some fifth-grade students exploring liquids—and discovering an unexpected outcome of their investigation.

Key Terms

sustained inquiry *(p. 153)*
cotyledon *(p. 162)*
germination *(p. 163)*
seed plant *(p. 166)*
angiosperm *(p. 166)*
taxonomy *(p. 166)*
kingdom *(p. 166)*
Monera *(p. 166)*
Protista *(p. 166)*
Fungi *(p. 166)*
Plantae *(p. 166)*
Animalia *(p. 166)*
photosynthesis *(p. 167)*
chlorophyll *(p. 167)*
monocot *(p. 169)*
dicot *(p. 169)*
vegetable *(p. 175)*
fruit *(p. 175)*
invertebrate *(p. 176)*
vertebrate *(p. 176)*
arthropod *(p. 182)*
univalve mollusk *(p. 182)*
bivalve mollusk *(p. 182)*

Extending Curriculum: Explorations of Liquids

FOCUSING QUESTIONS

- Which weighs more: a can of Coke or a can of Diet Coke?

- Where do you float better: on a lake or on an ocean?

- Why are science teachers often on the lookout for unexpected discoveries?

As students mature, their open-ended science investigations proceed in directions that we cannot always determine in advance. Sometimes the investigation leads to an unexpected exploration—and new science ideas.

As the teacher, you will always have thoughts and plans concerning the possible directions in which your students may take an exploration. Often, however, the students will surprise you, and the way you deal with the unexpected will have important implications for your students' learning.

In this chapter, we explore a story in which fifth-grade students examine the properties of liquids and pursue their own questions to higher levels of thinking and problem solving. The teacher is surprised that the students move toward inferences about the idea of density—it is not a topic she has planned to address—but she chooses to nurture their interests and help them expand their own thinking.

After seeing how the fifth graders proceed, we will look at a second-grade experience that seems much simpler in comparison. The second-grade activity addresses the basic ideas behind density by exposing students to objects that sink and float. You will see that such early experiences provide prior knowledge for the later, more sophisticated applications of the same science idea. Hence, the more varied and frequent that manipulative science experiences are in the early grades, the richer the possibilities are for students' prior knowledge and the better prepared they are for investigations in later grades.

As we explore these stories, think about your own role as a classroom teacher and how you will guide and facilitate experiences that allow students to move beyond the confines of a single experiment. Look for moments when the teachers:

- Acknowledge the students' ideas and reflect them back to the class.
- Help pull one experience together before going on to the next.
- Provoke students to look further.
- Build on their insights.
- Delight in the unexpected results of their explorations.
- Extend the lesson to pursue the new ideas and questions that arise.

SCIENCE STORY

Looking at Liquids

Ms. Drescher's fifth-grade class is in a small, urban community thirty-five miles north of a major metropolis. The students are from working-class and middle-class families with a diverse cross-section of ethnicities. Many of the students are immigrants and are struggling with English-language proficiency. Ms. Drescher arranges the students in heterogeneous groupings, mixed by race, ability, gender, and ethnicity. A spirit of cooperation pervades her classroom, and all the students are willing to contribute to investigations because they know they will have the opportunity to explore their own ideas. This elementary school goes from kindergarten through grade 5, so Ms. Drescher's students will be moving on to middle school next year.

A spirit of cooperation

The students are working on a physical science unit, studying the properties of liquids. They have learned that many kinds of liquids have several properties in common. Often, as the students have already seen, it is possible to notice differences among liquids as well. Today they are going to investigate the properties of yellow corn oil, clear corn syrup, and water to which blue food coloring has been added.

When the students return to their room after lunch, they find clear plastic cups and bottles of the three liquids. The blue-colored water is labeled as such. The labels on the bottles of corn syrup and corn oil are also in full view, and Ms. Drescher asks the students, "Where do you usually find these items?"

They respond with their own experiences of using corn oil and corn syrup, or watching others use them, to cook. Most of these students have seen corn oil used in the preparation of fried foods or in salad dressings. One student's family prepares plantains with corn oil. Only a few students have seen the corn syrup before. A couple of students know it is used in making candies, and one child's family uses it to make pecan pie.

Establishing connections

Ms. Drescher's thinking: *Ms. Drescher is eager to use materials that both work in the investigation and relate to the students' personal experiences. To establish the personal connections, she encourages a discussion of what the students already know about these liquids.*

Working in groups of three, the students examine cups of the three liquids. They have been asked to measure 150 milliliters of each liquid by using a graduated cylinder.

"What are we keeping the *same* about these liquids?" Ms. Drescher asks. Some children respond, "The same amount." Ms. Drescher looks at the class as a whole and says, "We know what scientists mean by 'the same amount.' Let's use the exact term." Several students then respond, "Volume."

Reinforcing terminology

Ms. Drescher's thinking: *Ms. Drescher is sure that the students know the meaning of* volume, *but she wants them to practice using the term.*

She holds up equal volumes of the three liquids and asks, "In what ways are these liquids the same? In what ways are they different?" The students share observations about color, thickness, and transparency.

Marisa offers, "You can see through the blue water and the corn oil, but it is harder to see through the corn syrup."

Another student notices that the cup of corn syrup feels heavier than the other cups. "Good observation," Ms. Drescher replies. "How can we prove that it is really heavier?" The students reply that they could use their balance scales.

How can we prove it?

Taking out their double-pan balances, the students observe that 150 milliliters (ml) of corn syrup weighs more than 150 ml of blue water or 150 ml of corn oil.

Ms. Drescher invites the students to think about what would happen if they combined the three liquids in one container. The students spend some time writing their ideas in their science journals. "To make a fair test," she asks, "what must we keep the same?" "The volume," the students respond. "Yes," replies Ms. Drescher, and she explains how important it is to keep the volume the same in combining the liquids.

Ms. Drescher's thinking: Ms. Drescher wants the students to see that "how much" you have of something can influence how it behaves with another substance. She knows that "doing science" means measuring and manipulating, but it also means planning investigations and thinking about procedures.

Ms. Drescher continues, "When we keep the amount or volume the same, we can honestly observe what happens when the liquids mix together. We can refer to the volume of the liquids as our *constant*, the property we keep the same. Then what is it that we are changing?"

The students respond that the liquids are different even though the volumes are the same. Ms. Drescher reminds the students that the things that change in an experiment are called the *variables*.

One student asks, "Which liquid should we pour first?" Ms. Drescher responds by saying, "That is your choice. Plan your investigation, and decide what volumes you will use."

Working in their groups, the students think about what will happen when the three liquids combine. They have a total of 150 ml of each liquid and lots of empty plastic cups. Some students think the corn oil will form a middle layer and the corn syrup will sink to the bottom; other groups predict that the oil will sink to the bottom. Some think the liquids will change as they mix together. All the students become involved in sharing their ideas and planning their investigations.

Making plans and predictions

Ms. Drescher's thinking: Because she wants the students to make choices about the plan of their experiment, Ms. Drescher does not provide lockstep directions. She is interested in seeing what variables they will use and which properties they will keep constant. How will they combine the liquids? What amounts will they begin with? Which ones will they pour first?

As it turns out, all the groups keep the volume of the three liquids the same by using the graduated cylinders. But the other procedures differ from group to group. Regardless of the methods they use, all the students are surprised when they combine their three liquids. "Oohs" and "ahs" can be heard as the syrup falls to the bottom, the oil rises to the top, and the blue water becomes the center layer.

Intriguing results

Many students are particularly surprised about the corn oil, and others are surprised that the results are the same regardless of the procedure. Ms. Drescher tells the students that they can spend the entire afternoon on their explorations. They write the results in their science journals. Some students also draw pictures of what they have observed.

Ms. Drescher's thinking: Ms. Drescher knew the students would need time to explore the liquids, so she planned the investigation on a day when she was certain that there would be no interruptions.

After a time, Ms. Drescher asks the students to sum up what they have found out as a result of their explorations. The students offer ideas, such as, "Some liquids float or sink in water." On a large poster pad at the front of the room, Ms. Drescher records their ideas, including the following comments:

> The liquids were different colors and different thicknesses.
>
> The oil and water layers are clear, and the syrup layer is only partly clear.
>
> For the syrup layer, only some light can pass through.
>
> The syrup layer poured more slowly than the other layers.
>
> The oil layer feels greasy.
>
> The syrup layer is the thickest and heaviest liquid.

"Well," Ms. Drescher continues, "let's reflect on what happened when you used the same amount of each liquid and weighed the liquids. Which was heaviest? Lightest?"

The students explain that the corn syrup was the heaviest and the corn oil the lightest. "Why do you think," she persists, "the corn syrup is always the heaviest?" The students reason that the corn syrup must have more "stuff" in it than either the water or the corn oil.

One student asks, "Can we do this again, only this time keep the *weight* of the liquids the same?" Noticing that the class seems enthusiastic about weight being the new constant, Ms. Drescher says, "Sure." She reminds the students that now that weight is their constant, they need to use their balance scales carefully.

As students begin to weigh their samples, they notice that the oil and water take up much more space than the syrup does for a given weight. That is because the corn syrup is so heavy; a little bit of it weighs the same as lots of corn oil or water. In other words, the students notice that the corn syrup has the smallest *volume* when the weights of the three liquids are the same. They wonder if it will still sink to the bottom even when there is so little of it.

They perform the investigation and—yes—the corn syrup always falls to the bottom, even though it has the smallest volume. They do not know why this has happened, and they turn to Ms. Drescher for help.

"Remember we said that the corn syrup sank to the bottom because it had more 'stuff'? Well," she says, "you have discovered another property of matter—the more 'stuff' in a substance, the *denser* it is."

"*Density,*" she goes on, "is how closely packed together the particles are." Some students then reason that the oil floats on water because it is not as dense as water and the water floats on the syrup because it is not as dense as the syrup. They are catching on to the idea that densities remain the same regardless of mass or volume.

Naming the new concept

Ms. Drescher's thinking: The idea of density was not in Ms. Drescher's plans for this experiment; she would have been content for the students to observe differences in weight among the liquids, along with the other properties they have already noticed. But since the students wanted to explore the relationship between volume and weight, she decided to welcome this unexpected outcome of their curiosity and expand on it.

The Next Day: Building Science Toys to Illustrate Density

The next day Ms. Drescher brings a number of decorative liquid-display toys to class— items that use liquids of different densities and colors to create movement and attractive illusions. You have probably seen these objects in stores. The students are surprised to learn that the liquids in these toys resemble the ones they have been working with. "That one looks like corn syrup and water," one student remarks.

Ms. Drescher then heats some corn oil gently by placing it in a heat-resistant glass pot on a warmer plate. She adds several drops of red food coloring, and the red drops seem to dance around in the warmed oil. "That looks like a lava lamp," one student says. Suddenly the entire class begins to guess the types of liquids contained in the other decorative objects.

Ms. Drescher then sets out for them a variety of clear plastic cups, Ziploc bags, glues, tapes, food coloring, and liquids. "Suppose you wanted to create your own science toy with these materials. What kinds of designs do you think you could make?" she asks.

THE MORE WE HELP CHILDREN TO HAVE THEIR WONDERFUL IDEAS AND TO FEEL GOOD ABOUT THEMSELVES FOR HAVING THEM, THE MORE LIKELY IT IS THAT THEY WILL SOME DAY HAPPEN UPON WONDERFUL IDEAS THAT NO ONE ELSE HAS HAPPENED UPON BEFORE.

—*Eleanor Duckworth (1996)*

FIGURE 9.1
· ·

A simple science toy using clear corn syrup and water tinted with blue food coloring.

As the corn syrup drips into the blue water, it settles to the bottom. Turning the cups over begins the process again, so that the syrup falls *through* the water to the bottom cup.

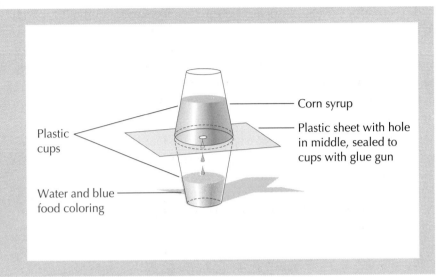

Plastic cups

Corn syrup

Plastic sheet with hole in middle, sealed to cups with glue gun

Water and blue food coloring

The students are excited at the possibilities, and Ms. Drescher asks them to work in their groups to develop their plans. "How will you create your own science toy?" she asks. "What liquids will you try? What volumes will you use? What do you want your toy to do?"

A new challenge

Each group is given time to explore ideas and plan a procedure. The students discuss their ideas—both scientific and aesthetic—with enthusiasm. Before the groups can begin building their toys, however, Ms. Drescher must approve their designs.

The groups discuss ideas . . .

Ms. Drescher's thinking: While giving the students the freedom to make their own plans, Ms. Drescher wants to be sure they are designing toys that can indeed be built with these materials and that the procedures will be safe.

Using the plastic cups and bags, and working in their groups of three, the students create their own science toys. With their understanding of which liquids will sink or float in water, they can create their own designs. Using tape and glue to fasten cups and bags together, they even have liquids flowing from one container to another (see Figure 9.1 for an example).

. . . and invent their own toys

E X P A N D I N G   M E A N I N G S

The Teaching Ideas Behind These Stories

■ Notice how Ms. Drescher prompts the students to make connections to their own lived experiences.

■ After actively listening to the students and inviting them to collaborate with her, she gives them reasonable choices.

■ To honor their ideas and encourage their experimentation, Ms. Drescher allows the students to repeat the experiment using mass as their constant. When their exploration leads toward a new concept, she revises her plans and brings in new materials so they can extend their investigation.

■ By inviting the students to design and construct their own liquid-display toys, she is giving them a chance to build on what they have learned.

■ The activities in which Ms. Drescher's students engage are pivotal to the science lesson, but they are not the entire lesson. She also provides students with opportunities to reflect on what they have discovered, talk about their ideas, and expand on their science experience.

The Science Ideas Behind These Stories

■ In elementary school science, students explore materials *within* categories and classifications as well as *between* them. Remember our earlier stories about solids, liquids, and gases (Chapter 7). In Ms. Drescher's class, the students are delving deeper into the category of liquids, realizing that substances in this category can be very different from one another.

■ By fifth grade, the basic question of what we keep the same and what we change becomes, "What are our constants; what are our variables?"

■ A **graduated cylinder** is a scientific measuring cup. It is a glass or plastic cylinder that is calibrated in milliliters for liquid volume—a handy tool in the classroom.

■ **Volume** is the amount of space an object takes up. Liquid volume is measured in milliliters and liters, and solid volume is measured in

TABLE 9.1   Densities of Some Common Materials

Material	Density (grams/cubic centimeter)
Cork	0.2
Wood (elm)	0.8
Alcohol	0.8
Olive oil	0.9
Water	1.0
Quartz	2.6
Aluminum	2.7
Iron	7.8
Nickel	8.9
Silver	10.5
Gold	19.3

cubic centimeters and cubic meters. When we refer to the *size* of an object, we usually mean its volume.

- **Density** is mathematically defined as the mass of an object divided by its volume. It is expressed numerically in grams per cubic centimeter. You can think of it as how closely packed together the particles are or as how many particles can fit into a given amount of space.

- A substance's mass or volume may vary, but the density of that particular substance—the relationship or ratio between its mass and volume—is always constant. The densities of some common materials are listed in Table 9.1.

- The concept of density is complex because it relates to two factors in a given material: its weight *as compared to* its volume. Comparing two variables requires careful planning and frequent measurement.

Questions for Further Exploration

- How does Ms. Drescher create an atmosphere of trust in her classroom?

- Think about the densities of some other common liquids: for instance, dish soap, red wine, milk, engine oil, olive oil, and house paint. How would they compare in density to corn oil and corn syrup?

Resources for Further Exploration

Electronic Resources

Chemistry Teaching Resources. **http://www.anachem.umu.se/eks/pointers. htm.** This site, sponsored by Umeå University in Sweden, includes links to various chemistry teaching resources, including curriculum material, software, and online journals. Though much of the material is for older grades, the site is also valuable for elementary school science.

Print Resources

National Science Resources Center. (1992). *Floating and Sinking. STC.* Burlington, NC: Carolina Biological Supply. Designed for fifth grade, these materials introduce students to a series of investigations with fresh water and salt water and their effect on buoyancy.

SCIENCE STORY

The Two Cakes

Having seen her class pursue the idea of density, Ms. Drescher decides to carry the subject further. A few days later she arrives at the classroom with two 9-inch-square baking pans. In one pan, she has baked a fluffy white angel food cake; in the other pan, she has baked a double-fudge brownie cake.

Following up on the students' ideas

"In what ways are these pans the same?" she asks.

"They both have cake in them," one student says.

"They are both the same size," another student remarks.

"They are both silvery looking," offers a third student.

Ms. Drescher then walks around the room and has several students feel the weight of each baking pan in their hands. "What do you notice when you hold these pans?" she asks.

The students respond that the brownie cake feels much heavier than the angel food cake. "So the cakes are the same size, but one is much heavier," remarks Ms. Drescher. "That means the brownie cake is denser than the angel food cake," blurts out one of the girls, making the connection to the earlier experiments.

The connection emerges

At that point Ms. Drescher cuts a small brownie and shows it to the class. "This brownie doesn't weigh more than the whole angel food cake in this pan," she notes. "That is because it is so small," one boy suggests. "If you had a piece of angel food cake the same size and then compared them," offers another student, "the brownie would weigh more."

Ms. Drescher cuts a piece of the angel food cake to exactly the same size as the brownie. They weigh the small cake pieces on a double-pan balance. The brownie does weigh more.

Proving the supposition

The students decide that density means, "If you have two objects exactly the same size, but one weighs more, the one that weighs more has the greater density."

The students' own definition

Ms. Drescher's thinking: Ms. Drescher believes that no definition of density could have more meaning for these fifth graders than this one—a meaning they have constructed on the basis of their own experiences.

SCIENCE STORY

The Cans of Soda

On another day Ms. Drescher brings in two soda cans, both unopened. One is Coca-Cola and the other is Diet Coke. She also has a large, deep, wide-mouthed jar, which is more than half filled with water.

"We're going to see what happens when we place these soda cans in the deep jar of water," she tells the class. But before performing the experiment, she asks the students to make predictions.

"They'll both sink. Soda cans are heavy," says Tiffany.

"They'll float," offers Justin, "because they have air bubbles inside of them."

Some students think the Diet Coke will sink and the regular Coke will float. Others believe the Diet Coke will float because it is "diet." They cannot tell just by holding the two cans of soda which one, if either, will float in water.

What do *you* think will happen?

After everyone has had a chance to think and predict, Ms. Drescher sets the cans in the water. The Diet Coke floats, and the regular Coke sinks.

Results of the test

The students try the experiment several different ways—they place the soda cans in the water in different order, on their sides, and upside down—but always with the same results. Ms. Drescher asks them to make further observations about the cans. They weigh the cans on double-pan balances and examine the data on the can labels. Here are some of their observations and inferences:

The class's observations and inferences . . .

The cans have the same volume: 335 milliliters.

The Coke can weighs more than the Diet Coke can.

The Coke has greater *density* than the Diet Coke—that is why it sank.

The Coke has 24 grams of sugar in it. The Diet Coke has no sugar in it.

The students want to know if the same thing would happen with other sugared and diet soft drinks. Ms. Drescher encourages them to bring in other soda cans and find out.

. . . and more questions!

Discovering what sinks and floats in water, children are grasping ideas about objects and their reaction to water. This is the foundation for more complex ideas about density in later grades.

Elizabeth Crews/Stock, Boston

The Floating Egg

The following week, Ms. Drescher brings a number of hard-boiled eggs to school with plastic spoons and containers of salt. She invites the students to explore what happens when they place a hard-boiled egg in a clear plastic cup three-quarters filled with water.

Another follow-up experiment

They try the procedure. "It sinks," one student quickly points out.

"Well, what does that mean? What happens when something *sinks* in water?" Ms. Drescher asks. The students decide that if an object falls to the bottom—or somewhere completely below the surface of the water—it has sunk. Ms. Drescher accepts their definition.

What does it mean?

This school is not far from the Atlantic Ocean, and Ms. Drescher is confident that students have had experience with salt water. "Let's see what happens to our egg," she says, "when we slowly add salt to the water and stir it to help it dissolve."

The students, working in pairs, slowly add the salt to their cups of water, first one teaspoon, then another. By the third teaspoon, the egg begins to rise; by the fifth teaspoon, it is floating on the water.

Changing one variable

"Okay," Ms. Drescher says, "let's talk about this. What's going on here? What is our constant?"

The students respond, "The egg."

"What is our variable?"

They call out, "The amount of salt in the water."

Ms. Drescher continues, "Okay, what did you notice?" The students start talking at once. Ms. Drescher lists their comments on the poster pad in front of the room:

More inferences and connections emerge.

The more salt, the better the egg floats.

The salt water makes the egg float.

The salt makes the water heavier.

Salt water helps you float better—that is why it is easier to float in the ocean than in a swimming pool.

The salt water is *denser* than the water from the tap.

The students reason that the salt dissolved in the tap water and gave the same amount of tap water many more particles. This made the tap water denser. The egg then was able to float on a liquid that was denser than the tap water.

One student asks, "Could the egg float on the corn syrup?"

"What do you think?" Ms. Drescher responds.

"Let's try it!" the students exclaim.

And they do. What do you think they discover?

E X P A N D I N G   M E A N I N G S

The Teaching Ideas Behind These Stories

■ Ms. Drescher keeps returning to the concept of density, offering the students a number of different ways to conceptualize it and see what it means in practice.

■ The students are engaged with very familiar materials: cake, soda, eggs.

■ All five of the major themes from the National Science Education Standards that we identified in Chapter 1 are embodied in this class's experience. The class had the opportunity to learn science by *doing,* by *inquiry,* by *collaboration,* over *time,* and by developing *personal knowledge.*

The Science Ideas Behind These Stories

■ The chemical additives used to sweeten diet drinks do not have the same mass as the sugar in regular sodas. It takes fewer grams of artificial sweetener to produce the same taste as 24 grams of sugar. That is why the mass of the Coke is greater than the mass of the Diet Coke.

■ When salt dissolves in water, the salt molecules separate or dissociate into charged atoms called *ions.* These ions fit between the molecules of water and make the same volume of water contain a larger number of particles. This increases the mass of the water but does not affect its volume.

■ Salt water therefore has a greater density than fresh or tap water.

■ Water exerts a lifting force on objects. This lifting force is called **buoyancy.** All objects appear to be lighter in water because of the buoyancy of water. Salt water exerts a greater lifting force than tap water does.

Measuring Water Displacement

Would you like to see how much water is displaced when a small object sinks or floats in water? Using three soft plastic cups and a graduated cylinder, you can perform your own test.

Take a plastic cup and cut a slit in it about 5 centimeters long and 2.5 centimeters wide. Fold this flap down like a spout. Then stand this cup on another cup that has been inverted. Place a smaller cup beneath the spout you have cut (see Figure 9.2).

Next, fill your cup with the flap until it overflows. Check that the overflow water is falling into the small cup; then discard this first bit of overflow water.

Your cup with the flap should now be perfectly full, up to the level of the spout. Gently place a small object in the cup. Take the displaced water that collects in your small cup and measure it with a graduated cylinder. Are you surprised by the amount?

- An object *floats* in water when any part of it is at or above the surface of the water. Objects *sink* when they are entirely submerged under the surface of the water.

- Objects that float in water displace water. For a floating object, the mass of the water displaced is equal to the mass of the object.

- Objects that sink in water also displace water. For an object that has sunk, the *volume* of water displaced is equal to the volume of the object. (See Figure 9.2.)

Questions for Further Exploration

- What do you think it would be like to learn science with Ms. Drescher?

- What is the actual numerical density of pure water? Salt water?

- What other foods or everyday materials would make good models of materials with different densities?

FIGURE 9.2   **An experiment in measuring water displacement.**   (a) Three plastic cups are arranged so that the one with a cut-out spout overflows into a collecting cup. To begin the experiment, the overflow cup is full and the collecting cup empty. (b) A small stone weighing 100 grams, with a volume of 40 cubic centimeters, is placed gently in the overflow cup. The stone sinks, causing water to overflow into the collecting cup. (c) The water from the collecting cup is measured in a graduated cylinder, indicating that 40 milliliters (cubic centimeters) of water were displaced. Thus, the volume of displaced water equals the volume of the stone.

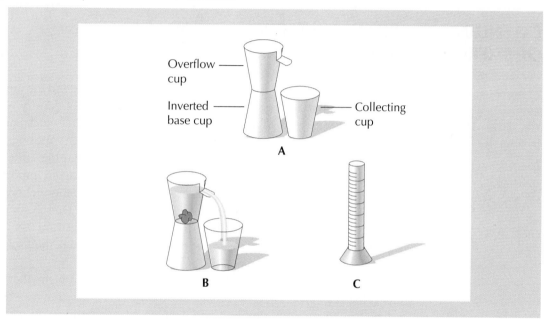

Resources for Further Exploration

Electronic Resources

Air Travelers. **http://www.omsi.edu/sln/air/.** This site, designed for upper elementary students, extends the concept of density to gases. It uses hot-air ballooning as an introduction to the basic principles of buoyancy, properties of gases, temperature, and the technology involved.

Buoyancy. **http://disney.com/DisneyTelevision/BillNye/.** This site is designed for children and teachers exploring specific topics in physical science, planetary science, and life science. Click on "U-Nye-Verse," followed by "Episode Guides" and "Physical Science" before selecting "Buoyancy." The explanation of sinking and floating and the exercise are excellent resources.

Print Resources

Buegler, M. (1988). *Discovering Density. GEMS.* Berkeley, CA: Lawrence Hall of Science. GEMS stands for Great Expectations in Mathematics and Science, a curriculum writing project from the Lawrence Hall of Science. To order GEMS materials, call (510) 642-7771.

EXTENDING CURRICULUM: TAKING ADVANTAGE OF EMERGING RELEVANCE

Why did Ms. Drescher carry on so long?

You may be wondering what prompted Ms. Drescher to extend the unit on liquids to such a degree. She not only seizes on the idea of density when the students bring it up, she carries on with it for days afterward. She offers multiple ways of approaching the idea, repetitions with variation, and new applications. We've seen this process in earlier chapters, but Ms. Drescher is a particularly powerful example. She has made an important pedagogical decision, electing to extend the unit in a direction where the students' natural curiosities are leading them. Clearly Ms. Drescher works in a school environment that encourages curriculum extension. As well, the students feel valued because their plans become central to their science lessons for several days.

We know that when students work on a science problem that has emerging relevance to them (Brooks & Brooks, 1993), they have a vested interest in solving the problem and a personal commitment to making sense of the solution. The term **emerging relevance** reminds us that relevance can *emerge* through teacher mediation, as it did in Ms. Drescher's class. It does not have to be preexisting.

Relevance can *emerge*.

What is clear is that the *students* must see the problem as relevant. You will know when questions or ideas of emerging relevance appear in your own classroom. The key is to start with science problems that make some connection to the students' lived experiences. These problems then become points of departure for other questions and other problems.

DON'T BE FRIGHTENED OFF BY COMPLEX QUESTIONS. TAKE THEM APART AND EXPLORE THEM.

—Eleanor Duckworth (1991)

Usually complex problems pose further questions to explore as students delve for deeper meaning. The three liquids that Ms. Drescher used, layering as they do, are complex enough to promote further exploration. Remember, as we saw in Chapter 7, that oversimplification is not in the students' best interests. In fact, it often leaves them feeling more confused than complex problem solving. So don't be frightened off by complex questions. Take them apart, explore them with the students, and guide the students to consider possible experiments. If Ms. Drescher had closed off the explorations of the three liquids, a complex concept like density would have been lost to her fifth graders at a time when it was relevant to their explorations.

Your decision to pursue emerging problems should be based on a number of factors:

Factors to consider

- The availability of materials.
- The relevance of the emerging problem to the overall objectives of your science teaching.
- The ways in which the students see the exploration as relevant to *them*.
- The students' investment in wanting to know.
- What you have learned about their facility in pursuing investigations.
- Your own belief that this extension will strengthen their understanding of the natural world.

Allowing enough time

As a general rule, it is a good idea to allow more time for a science lesson than you may initially think it requires. The process of exploration takes time, and the follow-up experiences take time as well. Also, make sure that your science experiences include relatively inexpensive materials that are easily accessible. It is hard to extend the curriculum if you cannot supply the right materials.

In the next story, we visit a second-grade class in the same school district. Notice how early ideas about size and materials influence the discussion. Notice as well how the teacher poses a relevant science problem. As you read, think about how an experience like this in second grade relates to later activities like the ones we saw in Ms. Drescher's fifth-grade class.

SCIENCE STORY

Floating and Sinking Fruits

It is the week of Halloween, and Ms. Sacco's second-grade classroom is filled with items pertinent to the season, including pumpkins of varying sizes. The students are exploring the properties of pumpkins.

They measure the pumpkins with string, holding the string around the fattest part of each pumpkin and then marking the string to indicate this distance. Then they lay out the string against the calibrated centimeters of

Measuring pumpkins

a meter stick. With this procedure, they find out how many centimeters around each pumpkin is at its bulging middle. They learn that this measurement is called the *circumference* of the pumpkin, and it may be compared to the equator of the earth. Ms. Sacco holds up a globe and shows the students the equator.

The children are excited about the season. They know they are going to open their largest pumpkin and carve a face in it. First, however, Ms. Sacco asks them to make as many observations as possible about this pumpkin, and she lists them on a poster board:

Initial observations

It is large.	It has no smell.
It is round.	It has a stem.
It is orange.	It has lines on it.
It is hard.	Its circumference is 60 centimeters.

Ms. Sacco asks, "What do you think will happen if we place our pumpkin in a basin of water?" The children laugh. "It will sink, of course," they reply.

The children's prediction

Ms. Sacco takes the class to the large basin in the custodian's work area, and they hover around her as she fills the basin with water. As she places the pumpkin in the water, the children see that it bobs and floats! Squeals of surprise and chatter ring out as they walk back to their classroom.

Surprising results

The students do not know why the pumpkin floats. Ms. Sacco states, "Let's open the pumpkin, explore its insides, and do some more experiments to solve our mystery."

She cuts out a lid around the stem of the pumpkin, and the children look inside. They see a large cavity and, within it, the pulp—the seeds and mushy, moist connecting threads.

Looking for a reason

"What else is in there?" Ms. Sacco asks.

"Nothing," the children reply.

The next day Ms. Sacco brings grapes, oranges, apples, and a green pepper to class. "In what ways are these foods all the same?" Ms. Sacco asks. The second graders respond:

They are sort of round.

You can eat them.

They have skin.

They all come from plants.

Then the students use a large basin of water to explore whether these items will sink or float. They are shocked to find that only the grape sinks.

Testing similar objects

Science ideas emerge when students explore pumpkins. What's inside? How many seeds? Does it sink? Float? How can we measure it? Let's plant the seeds. Is there a sink large enough to float that pumpkin?

Lawrence Migdale

All the other foods float. Michael wonders why a little grape will sink, but a great big pumpkin floats. It is time to do more investigating.

Ms. Sacco peels the orange and the grapefruit. Now they both sink!

The children explore the skins of these fruits with a magnifying lens. When they hold the skin up to the light, they notice spaces.

"What do you think is in these spaces?" Ms. Sacco asks. Some students say "nothing"; others say "air."

An idea begins to develop.

Ms. Sacco invites them to look for what they think might be air spaces in the other fruits. The students find that the floating apples have spaces in the center cavity where the seeds rest. The green pepper and the pumpkin have large empty spaces. The orange has the spaces in its peel. "The orange peel was the life jacket for the orange," Roselia says, and the other students laugh.

Ms. Sacco's thinking: In an earlier lesson on states of matter, the students explored plastic bags with air inside them. Still, Ms. Sacco has noticed, many students are reluctant to acknowledge that air is inside the spaces in the foods. She is hoping to guide them to this idea. She also hopes that the students will notice that the size of the fruit cannot tell them if it will sink or float in water. Size (or volume, as it is later called) is less important than what is inside the object.

To pull the lesson together, Ms. Sacco invites the students to think about what makes some fruits float and others sink. She distributes grapes and magnifying lenses, and the children cut the grapes in half with plastic knives. They observe that there seem to be no open spaces inside a grape. That is why it sinks, they decide. They conclude that a fruit probably will float if there are open spaces inside it.

The students reach their own conclusions.

E X P A N D I N G   M E A N I N G S

The Teaching Ideas Behind This Story

- To help the children make connections between science and their world, Ms. Sacco is using seasonally relevant materials to explore nature.

- Notice how she challenges the students to draw comparisons and consider properties of the various fruits.

- Ms. Sacco makes connections between the pumpkin and our planet, paving the way for the notion of using different objects as models for the earth (see Chapter 10).

- The students use mathematical skills through measurement of the circumference.

- When the students do not know why the pumpkin floats, Ms. Sacco does not tell them. She sanctions the question and gives the students confidence to believe that they will find the answer. She then provides additional experiences that help them do so.

- Ms. Sacco avoids emphasizing that "air" is inside the fruits that float.

The Science Ideas Behind This Story

■ Some fruits float in water because they have air inside them. The air decreases the overall density of the fruit, making it less dense than water.

■ Objects with a density less than the average density of water will float on water. The average density of water is 1 gram per cubic centimeter. This means that 1 cubic centimeter of water (equal to a milliliter of water) has a mass of 1 gram.

Questions for Further Exploration

■ If iron has a density of 7.8 grams per cubic centimeter, why do steel ships float?

■ What is the ratio between the grape's mass and volume? Greater than 1? Less than 1?

■ With what other common materials could children do the sink-or-float test?

Resources for Further Exploration

Electronic Resources

Science and Mathematics Initiative for Learning Enhancement (SMILE). **http://www.iit.edu/~smile/index.html.** This is a collection of elementary and middle-grade lessons in chemistry, biology, physics, and earth science from teachers all over the country. Its chemistry link, **http://www.iit.edu/~smile/cheminde.html,** has several lessons on density, including density of liquids.

Print Resources

National Science Resources Center. (1993). *Balancing and Weighing STC.* Burlington, NC: Carolina Biological Supply. This is a series of lessons designed for the younger grades to explore relationships involving balance, weight, and size. Students work with cupfuls of food. For information, call (800) 334-5551.

LOOKING BACK TO LOOK AHEAD

You may be wondering why we visited Ms. Sacco's second-grade class after exploring Ms. Drescher's fifth graders at work with volume, mass, and density. The second graders were simply exploring sinking and floating fruits. Where's the connection?

Remember the discussion in Chapter 3 of the process by which children construct meaning. Our discussion was influenced by the learning theory called constructivism and its implications for teaching. Far from being a neat, linear process, the construction of new ideas is based on a recursive process of visiting and revisiting prior conceptions, altering our views based on new experiences, reflecting on those views in peer groups, and then, sometimes, formulating a new idea.

Constructing ideas—a complex, recursive process

What takes place in Ms. Sacco's class has important implications for what students can learn from the experiences in Ms. Drescher's class. That is, if Ms. Drescher's students have had prior experiences like those provided by Ms. Sacco, they will be better prepared to understand concepts like volume and density. They can use that prior knowledge to build their new, more complicated ideas.

A gradual building of understanding

The children in Ms. Sacco's class believe that the large pumpkin will sink. They find out that that belief is not accurate, and they explore further. This work leads to a partial understanding that the materials inside an object determine one of its properties. They begin to understand that some objects have less "space" inside than others, and this property can be more important, for some purposes, than the size of the object. The students are approaching one conception of density—how closely packed the particles are. By the time they have a full-fledged lesson on density, in Ms. Drescher's class or elsewhere, they will be ready for it.

In the next two chapters, as we explore additional activities for the upper elementary grades, think about the prior experiences that help the students construct new ideas.

Key Terms

graduated cylinder *(p. 193)*
volume *(p. 193)*
density *(p. 194)*
buoyancy *(p. 199)*
emerging relevance *(p. 202)*

10 Making Models: Explorations of the Solar System

FOCUSING QUESTIONS

- What models have you made? Did you ever build a model airplane? A clay model? An abstract model?

- What objects of nature can be explored only through a model?

- How can making models facilitate learning?

When I was in fifth grade, I made a model of a tooth for Dental Health Week in my elementary school. In order to build it, I needed to gather pictures of teeth and understand the various parts of a tooth. I had to decide which type of tooth to represent and figure out the size and composition of the model. I decided on a molar tooth (molars are the grinding teeth toward the rear of your mouth) and plaster of Paris for the material. I chose to make the model 8 inches tall, reasoning that it would be large enough to see and that I could draw lines on it to represent the insides.

What I remember most from making that tooth model—aside from my struggle to create the plaster mold out of clay—was that, to my surprise, the tooth is a complicated structure with lots of layers. In daily life we see only the crown, the visible part of a tooth. My model, though, included markings to show where the root canal, the nerve, and the pulp are.

Although not exact by any means, my model tooth remained on display for all of Dental Health Week. Moreover, I never *forgot* the structure of a tooth. Decades later, when I needed to have the contents of the pulp chamber of one of my molars removed because of infection, I pictured the root canal from my model.

This chapter explores what happens when we engage students in making models of natural objects. I think you'll see that your students can have experiences as instructive and memorable as mine. We're going to look at models of the solar system. First, though, we need to answer the question: What is the point of making models?

THE USEFULNESS OF MODELS

The activities you have been reading about in earlier chapters invite students to become directly involved with their objects of study. Sometimes, however, it is not possible to explore the objects directly. They are too large or too small, too far away, or inaccessible for other reasons.

Physical models

When scientists are unable to work directly with materials, they construct models of the materials in an effort to gain a greater understanding of their structure and function. A **model** may be a physical structure, either a smaller or a larger representation of a system or an object. My model tooth was one such structure. Scientists have long made physical models with whatever materials are available to them. One of the most famous involved the chemical structure of deoxyribonucleic acid, known as DNA— the material in our cells that passes genetic information from one generation to the next. The model developed by James Watson and Francis Crick—one of the most important scientific revolutions of the twentieth century—involved materials resembling giant tinker toys.

Scientists construct models after they have gathered enough data about their objects of study to begin to make a reasonable facsimile. In the case of DNA, the scientists used data from an X-ray crystallography method employed at the time by Rosalind Franklin. Dr. Franklin was able to gather images of patterns made by the DNA molecule when it reflected X-rays. Her data became the basis for Watson and Crick's DNA model.

Mental models

A model does not have to be something you can touch. It may be a mental construct—a design that forms an image in your mind representing a concrete process or object. For example, I carry a mental model of an atom in my mind; it is the way I have conceptualized an atom on the basis of what I have learned.

Computer models

Models can also be computer programs or computer-generated images. Today, a great deal of scientific research is performed through the use of computer models. As just one example, meteorologists often explore potential weather events by using computer simulation based on actual

James Watson and Francis Crick in 1951, showing their Nobel Prize winning model of the genetic material, deoxyribonucleic acid (DNA). Making models is one way we learn about nature.

A. Barrington Brown/Science Source/Photo Researchers, Inc.

satellite data. You see the results of these computer models on the television news each day when forecasters describe the possible tracks a storm system may take.

As you will discover from the following science stories, making models is a way of furthering your students' understandings of objects and events that they cannot manipulate directly. The nature of the scientific investigation changes somewhat when we need to gather data in order to construct models. In previous chapters, you saw the data gathering being done while the students manipulated their objects of study. In this chapter, students gather research data first, *then* construct the model, *then* look for deeper meanings. Watch what happens when we visit Mr. Johnston and a fourth-grade class as they build and explore a model of the solar system.

SCIENCE STORY

An Edible Solar System

Mr. Johnston is a veteran elementary school teacher. He has a science room in a small, suburban elementary school, and students visit his room once a week with their classroom teacher. Together the students engage in science activities, which they continue in their regular classroom. There are interesting materials on display and arranged in storage cabinets all around the room. Three hamsters, a rabbit, and numerous plants share the room with Mr. Johnston.

This particular day, when fourth graders arrive for their morning visit, the only materials Mr. Johnston has set out are round fruits and vegetables of varying sizes: grapes and peas and cabbages and grapefruits, melons of different types, apples of varying sizes, oranges, apricots, and small tomatoes.

Familiar materials

"Are we dissecting fruits and vegetables again?" the students ask.

Mr. Johnston laughs. "No," he says. "We are going to use these fruits and vegetables to make a model of the planets in our solar system."

The students have been exploring "objects in the sky," and they have just completed a huge poster-board model of the sun, so this project makes sense to them. They are excited to begin. "How do we do it?" they ask.

Mr. Johnston explains that they need to gather a lot of information before they can use these materials to make a model. "Remember all the research we did on the sun?" he asks. "What type of information do we need about the solar system in order to construct a reasonable classroom model?"

The students brainstorm various questions to answer, including the following:

The students' list of questions

What are all the objects in the solar system?

How far away from Earth are the rest of the planets?

How many moons does each planet have?

What are the planets made of?

How many planets have atmospheres?

How big are the planets?

We construct models in order to make meaning of objects that are too remote and impossible to study through direct manipulation. Students created this model of Saturn to help conceptualize their information about the planet.

Anthony Freeman/Photo Edit

How far away from the sun is each planet?

What colors are the planets?

Mr. Johnston records the questions on poster paper as the students record them in their science journals. He invites the students to work in groups of four and to decide on the particular questions they want to research. He explains that they are in the "data-gathering" phase of this model-making project, and they must select the questions they are most interested in exploring.

Mr. Johnston's thinking: Typically, solar system models are constructed from Styrofoam balls of varying sizes. The assumption people may make when observing this type of model is that all the planets are the same except for their size. That is not true. Planets differ from each other not only in size but also in composition and surface features. For that reason, Mr. Johnston thinks that fruits and vegetables will make a better representation, as well as a more interesting one.

Why use these materials?

After selecting their questions to research, some groups use the resource books available in their science room; others go to the school library; still others access the Internet. They accumulate a good deal of information about the planets individually and the solar system as a whole.

Gathering information

One critical piece of information, all the students decide, is the planets' distances from the sun. Here Mr. Johnston intervenes, explaining that the distances between the planets and from the planets to the sun are so huge that no classroom model can be truly accurate for distance. He tells the students that one fourth-grade teacher, in an effort to represent the distances accurately, spread out the "planets" all over the local community (Whitney, 1995).

Mr. Johnston's thinking: Mr. Johnston knows that the fruit-and-vegetable model will not be completely accurate, but that is okay, especially if the students come to understand how it resembles the real solar system and how it differs. Later, he will ask them to make comparisons between the model and the real thing.

> **GIVEN A THREE METER SUN AT ONE END OF THE [SCHOOL] BUILDING, THE SCALE MODEL PLANETS WE CREATED WOULDN'T FIT IN THE SCHOOL. IN FACT, PLUTO WOULD BE NEARLY 15 KM AWAY.**
>
> —David Whitney

To make the calculations more manageable, Mr. Johnston provides the students with distance dimensions in the form of astronomical units. One astronomical unit (AU) equals 93 million miles, the average distance from the sun to the Earth. All the other distances in AUs are relative to that distance. Table 10.1 lists these distances.

Mr. Johnston distributes calculators to help the students explore the numbers. He uses an overhead projector calculator, while the students use their own calculators at their tables. Together, they multiply each of the distances in AUs by 93 million miles. The enormity of the numbers indicates to the class how huge the solar system really is.

Now the students turn to the data they have collected on planet diameters, given in kilometers (see Table 10.1). Using these data, the students order the planets from smallest to largest. Then they are ready to select the fruits and vegetables that will be most representative of the solar system.

Exploring the data

TABLE 10.1   The Solar System Data Used by Mr. Johnston's Class

Planet	Average Distance from Sun (AUs)	Diameter (km)
Mercury	0.4	4,878
Venus	0.7	12,104
Earth	1.0	12,756
Mars	1.5	6,796
Jupiter	5.2	142,796
Saturn	9.5	120,300
Uranus	19.0	52,400
Neptune	30.0	48,600
Pluto	39.0	4,000

For the sun, they choose a huge pumpkin that Mr. Johnston has brought in. In reality, the sun's diameter is about 109 times that of Earth, and if the sun were a hollow ball, one million Earths would fit inside. Thus the students know that the pumpkin is not really big enough to represent the sun in their model. As for the planets, different groups of students make different decisions. Here are the choices one group makes:

One group's choices for the model

Mercury: a small cherry tomato

Venus: an orange

Earth: an apple, slightly larger than the orange

Mars: an apricot, approximately half the size of the apple

Jupiter: a honeydew melon

Saturn: a cantaloupe

Uranus: a cabbage

Neptune: a grapefruit, slightly smaller than the cabbage

Pluto: a tiny pea

For many students, it becomes a challenge to remember which fruit is modeling which planet, so Mr. Johnston instructs each group to make a key.

Once the students have made their selections, it is time to go outside and place the model planets in such a way that they also make a model of relative distances from each other and the sun. The class chooses the set of vegetables and fruits listed above as the ones to take outside. Mr. Johnston suggests using 1 meter to represent 1 AU. (This is about the span of a fourth grader's "giant step.") With this technique, the students observe that the first four planets occupy about 1½ meters of space, while Pluto is 39 meters away from the designated sun. The student carrying the little pea model of Pluto finds it difficult to see the student carrying the huge pumpkin model of the sun (see Figure 10.1).

Modeling distances

Back inside, when the students discuss their model, Mr. Johnston asks, "In what ways is our model like the real solar system?" The students offer many responses, including these:

Model and reality: comparisons . . .

"All the planets are different, with different textures and insides."

"The relative sizes of the planets are the same, more or less."

"We set them up so the relative distances are similar."

FIGURE 10.1   An edible solar system model created by Mr. Johnston's class.

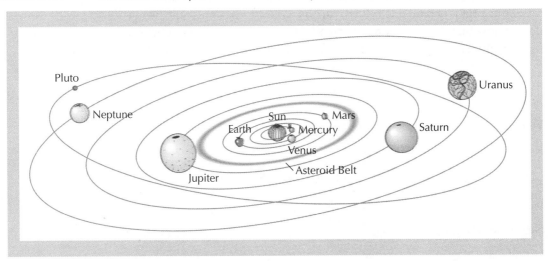

"They're in the order of their distance from the sun."

"In what ways is the model *different* from the real solar system?" Mr. Johnston then asks.

"You can't eat the real planets!" the students shout, laughing. And they add other differences as well:

. . . and contrasts

"The model is so much smaller than the real thing."

"The planets aren't moving."

"The distances are much much larger."

"Some planets have moons."

Mr. Johnston's thinking: The students are doing a good job in evaluating this particular model, Mr. Johnston believes, and they are beginning to learn a crucial fact about models in general: you always have to be aware of how your model differs from the reality it represents.

A crucial fact about models

The remark about planets' having moons leads to an interest in adding moons to the model. On another day, the class decides to use miniature marshmallows and toothpicks to represent the moons of Earth (1), Mars (2), Jupiter (16), Saturn (17), Uranus (5), Neptune (3), and Pluto (1). This activity generates further questions. For example, one group wonders if the

Extending the model

minimarshmallow moons are on a similar scale to the planets. They do some research on the Earth's moons and the moons of Jupiter, and they realize that a regular-size marshmallow would be a better model for our moon, which is about one-fifth the size of the Earth.

To conclude the lesson, Mr. Johnston asks the students to observe their models closely, think about what they have learned, and write their conclusions in their science journals. He asks them to be sure to answer these questions:

Mr. Johnston's questions

What planets seem similar in size to one another?

What planets are Earth's neighbors in the solar system?

What place in the solar system is the Earth in order of distance from the sun? In order of size?

Which is your favorite planet to eat?

The students finish the session by eating their favorite parts of the model. "Yum," one fourth grader says. "I just love Mars!"

EXPANDING MEANINGS

. . . .

The Teaching Ideas Behind This Story

- Inviting the students to generate their own questions about the solar system is a first step toward giving them ownership of the process of gathering the data.

- Notice how Mr. Johnston encourages students to discover both the accuracies of their model and its limitations.

- This experience of gathering data and creating an edible solar system model is directed toward students in fourth grade and beyond. You can adapt the activity for students in earlier grades. Instead of numerical data, you can use pre-cut circles to represent relative sizes of the planets. Students can select their fruits and vegetables using these circles as their data. In the younger grades, each astronomical unit becomes one "giant step."

■ It is the teacher's job to select a variety of fruits and vegetables that work. Using your data, choose carefully. Obviously fruits are not of uniform size, so your selections will be based on the season and the fruits that are available where you live.

The Science Ideas Behind This Story

■ Our **solar system** is made up of a group of heavenly bodies that move around the sun. The main members of the solar system are the nine **planets.** If you cannot take your students outside as Mr. Johnston did, you can use Table 10.2 to construct a model of planetary distances inside the classroom.

■ Between Mars and Jupiter is a belt of several thousand **asteroids** of different sizes. These are like tiny chinks of planet, and they also move around the sun.

■ The sun is the only member of the solar system that is a star.

■ **Stars** shine by producing their own light.

■ Planets shine by reflecting the light of the sun or of other stars.

■ All planets travel in their own **orbits** (closed paths) around the sun, moving counterclockwise around the sun, from west to east.

TABLE 10.2   Relative Planetary Distances Expressed in Units Suitable for a Classroom Model

Planet	Distance from the Point Representing the Sun
Mercury	1.75 inches
Venus	3.25 inches
Earth	4.75 inches
Mars	7.0 inches
Jupiter	2 feet
Saturn	3 feet, 8 inches
Uranus	7 feet, 5 inches
Neptune	11 feet, 8 inches
Pluto	15 feet, 3 inches

- The solar system is an obvious example of materials that we cannot manipulate directly. While some objects in the sky may be directly observed, most parts of the solar system need to be researched in libraries or on the Internet.

Questions for Further Exploration

- What other round objects with diverse colors and textures could model the solar system?

- Suppose you could not find a large pumpkin. What else could you use to model the sun?

- What materials could represent the belt of asteroids?

- How does the sun make its own light?

Resources for Further Exploration

Electronic Resources

Arnett, Bill. *The Nine Planets: A Multimedia Tour of the Solar System.* **http://seds.lpl.arizona.edu/nineplanets/nineplanets/nineplanets.html**. Excellent for elementary students, this site describes current knowledge about each of the planets and moons in our solar system.

Print Resources

Council for Elementary Science International. (CESI). (1991). *Water Stones and Fossil Bones.* CESI Sourcebook VI. Washington, DC: National Science Teachers Association and CESI. This volume has 51 illustrated science activities by many authors, including a model solar system activity that is an extension of the edible model.

Sutter, D., Sneider, C., & Gould, A. (1993). *The Moons of Jupiter. GEMS.* Berkeley, CA: Lawrence Hall of Science. Another fine series of suggested lessons from GEMS. In these, students are introduced to the work of Galileo and other astronomers. Students create a scale model of Jupiter using their schoolyard, and they explore photographs of Jupiter's moons from the *Voyager* spacecraft.

MODELS AND MEANING

I have often built an edible solar system with elementary school students, and they tend to remember the project for years to come. The first time was in my daughter's third-grade class. At her high school graduation, a former third-grade classmate of hers greeted me and said, "I remember when you visited our class and the pumpkin was the sun."

From model to meaning

The fruits and vegetables make a very useful model for all students, especially for those who need to work with concrete objects to shape comparisons and interpret data. But remember that the construction of this model was not an end in itself. It was a step toward a meaningful understanding of the objects in the solar system. In Mr. Johnston's class, using the materials generated further research, as when one group wondered about the sizes of moons and decided to research the topic.

Making a model and interpreting data based on the model, exploring comparisons between the model and the real object, moving the model as though it were the real thing—these types of activities have important im-

Preparing for future learning

plications for future learning. Besides facilitating the understanding of abstract concepts, the use of models helps prepare students for the science concepts they will encounter in later grades. For example, balancing chemical equations involves using chemical symbols and a mathematical process to model the actions of real atoms and molecules.

In the next science story, Mr. Johnston's fourth graders explore the properties of the planets' orbital paths around the sun. This lesson requires another type of model making.

SCIENCE STORY

A Model Orbit

On another day, Mr. Johnston distributes string, pencils, centimeter rulers, and pushpins to the class. He asks the students to work with a partner and explains that they are going to draw a shape that represents the path of planets around the sun.

First, he has each pair of students tie a loop with a string about 30 cen-   **The set-up**
timeters long. Then he invites them to insert two pushpins toward the

middle of one of their journal pages, placing them about 8 centimeters apart, and making sure the pins go through several pages. The pins should be anchored securely, Mr. Johnston explains. He goes around the room and supervises as each pair of students sets up the pushpins.

Now the students are ready to proceed. They place the loop of string over the pushpins. Then, anchoring a pencil in the loop and keeping the loop taut, they trace a figure that resembles an oval (see Figure 10.2). Each team of students compares its shape with those drawn by others. Indeed, the shape is always oval, no matter who draws it.

The experiment

Mr. Johnston tells them that another word for this shape is **ellipse,** and he asks them to label the two points where their pushpins were inserted the **foci** of the ellipse. "In what way is this image different from a circle?" he then asks.

New terms—why doesn't the teacher define them?

The students say that "you can draw a circle with just one pushpin." Also, they remark, a circle has a center, but the oval does not have one center.

Mr. Johnston explains that the planets travel in an elliptical orbit around the sun. We can think of the sun, he says, as located at either one of the foci of the ellipse.

Mr. Johnston has also provided three large pieces of foam board—thick, white pieces that he has hung from hooks at the top of the chalkboard. Each piece is about 1 meter square. He invites students to come up and place pushpins at different distances. Then he suggests that the students use string loops of varying lengths to determine in what ways the shape of the ellipse changes when the length of the string and the distance between the foci change. (For safety, Mr. Johnston distributes the pins only when the students come up to the board. He asks them not to walk with their pushpins.)

Modifying variables

FIGURE 10.2
. . . .

Drawing an ellipse (oval) with a pencil, string, and two pushpins.

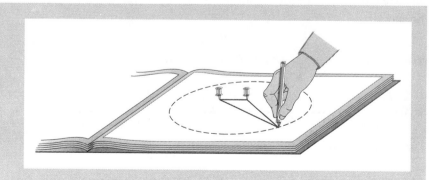

"Be sure to measure the distance between the pins and the length of the string you use in centimeters," he reminds the students. "Record your data in your science journals." He watches as the students draw their different ellipses on the foam boards.

Mr. Johnston's thinking: Mr. Johnston knows that their new pencil tracings will change in shape as the length of the string and the distances between the pins increase. The challenge for the students will be to make the leap from these pencil images to the changes in the orbital path of a planet when its distance from the sun increases.

The students do notice how exaggerated the elliptical shape becomes as they increase the distance between the pins. Mr. Johnston suggests that they keep the distance between the two pushpins constant, changing only the length of the string, and they try the experiment in that way.

The students' observations

"Now, let's imagine that any one of these ellipses represents the orbital path of the Earth around the sun," Mr. Johnston suggests. He labels one of the foci with an "S," for the sun. "Let's draw a line representing the distance from the sun to the point on the ellipse where the Earth would be closest to the sun and a line to the point on the ellipse where the Earth would be farthest from the sun." Mr. Johnston draws the lines with a ruler as the students watch.

Exploring features of the model

Mr. Johnston's thinking: Since the orbital paths of the planets are ellipses, there must always be a point where a planet is closest to the sun and another point when it is farthest from the sun. Mr. Johnston is hoping that the students will see this.

Next, Mr. Johnston has the students observe pictures and drawings of the solar system from resource books and the Internet. Using these resources as well as their own drawings of the ellipses, the students make inferences about the time it takes for each planet to travel once around the sun. They reason that Pluto's journey must be the longest and Mercury's the shortest.

Extending the research

E X P A N D I N G   M E A N I N G S

The Teaching Ideas Behind This Story

■ Notice how Mr. Johnston engages the students in drawing first the same ellipses and then all different sizes.

■ Mr. Johnston offers labels for their images without precisely defining the terms *ellipse* and *foci*. The students will develop those definitions from their experience of drawing their figures.

■ Mr. Johnston uses each opportunity to extend the students' thinking about the solar system. The ellipse drawings become a springboard for further research about the time it takes for planets to complete each revolution around the sun.

The Science Ideas Behind This Story

■ A **satellite** is any heavenly body that travels around another heavenly body. The planets are satellites of the sun. The moon is a satellite of the Earth.

■ The planets' orbits around the sun are elliptical (oval).

■ An ellipse has no center. It has two foci or focus points. The sun is located in space at one focus of the planetary ellipses.

■ The point in a planet's orbit when it is closest to the sun is called its **perihelion.**

■ The point in a planet's orbit when it is farthest from the sun is called its **aphelion.**

■ The time needed for a planet to make one complete turn or revolution about the sun is called its **year.**

Questions for Further Exploration

■ What is the difference between a center and a focus?

■ What are some implications of the planets' changing distances as they orbit the sun?

■ Do you think planets travel at uniform speed in their orbits?

■ If you were a Martian, how old would you be? That is, what is your age in Mars years?

Resources for Further Exploration

Electronic Resources

NASA Spacelink. **http://www.spacelink.nasa.gov/index.html**. Sponsored by the National Aeronautics and Space Administration, this site describes space travel projects and more, with excellent graphics and links to instructional materials.

Print Resources

VanCleave, J. P. (1991). *Astronomy for Every Kid*. New York: Wiley. Wonderful activities in this book address different properties of the planets, planetary motion, and the moon.

Whitney, D. (1995). The case of the misplaced planets. *Science and Children,* 32(5):12–14. Whitney, a fourth-grade teacher, tells of his discovery that, given a 3-meter sun, his students' scale model of the solar system required that some planets be placed outside their town.

SCIENCE STORY

Shapes of the Moon

In the early autumn, the students in a fifth-grade class in rural Maine have been exploring models of the solar system, and they have shown an interest in learning how the moon, the Earth's satellite, appears different at different times of the month. One day their teacher, Ms. Cairo, tells them that each of them will begin keeping a *moon-phase journal*. It will be an opportunity to observe part of the solar system directly, outside the classroom.

In this journal, she explains, they should record their observations of the moon over a period of six weeks. She gives them these general instructions: Ms. Cairo's instructions

Search the sky on a daily basis.

Keep daily records of your attempts to see the moon—whether you see it or not.

When you do see it, include the following things in your journal:

(a) A description or drawing of what you saw.

(b) The time and the date.

(c) What you were doing at the time.

(d) Anything else you want to write down.

Become more aware of when the moon is visible in the sky.

Watch for changes in the moon's apparent shape, and try to figure out the sequence of these changes.

Ms. Cairo's thinking: Ms. Cairo believes the students can best learn about the moon's phases through direct research. But she wants to leave the precise format of the moon-phase journal open for the students themselves to decide. This open format, she thinks, will be especially useful for students who express themselves better through drawing than they do verbally.

Over the next six weeks, the students observe the sky and record their data in both words and pictures. Once a week, Ms. Cairo checks their progress and inquires about any difficulties they may be having. At the

Six weeks of journal writing

Denise's art talent emerges as she prepares her images for her moon phase journal.

Denise Monda

end of the six weeks, she asks them to bring their journals to class to share their observations.

Excited, they arrive in the morning to find large sheets of paper taped to the chalkboards in the room. On each sheet a heading reads, "WHAT WE FOUND OUT ABOUT THE MOON." The students take markers and begin filling in the sheets with their observations, discussing them as they do so.

Ms. Cairo's thinking: There are many ways to encourage the students to share their observations. This method, Ms. Cairo thinks, will allow students who are not always comfortable speaking publicly to display their work.

Here are some of the observations that the students write down for the class to see:

- Sometimes, when the moon is rising or setting, it looks yellowish or even orange.

- The moon looks bigger when it is rising or setting than when it is high in the sky.

- You can see the moon in the early morning and the late afternoon at some times of the month.

- You can't see the moon when the sky is very bright in the middle of the day.

- Sometimes you can see a very pale full moon in the sky early in the morning.

- Sometimes, when there is a crescent moon—appearing as a sliver in the sky—you can see the rest of the moon faintly lighted.

- After the crescent moon gets bigger, you can see a half-circle moon in the sky.

- You can see this half-circle moon on the way home from school.

- The moon gets bigger as you watch it until it gets to a whole circle—a full moon.

- You see the full moon at night, but not in the morning.

- The moon gets smaller each night after the full moon.

- When the moon is almost full, it is bulging on one side. Then when it starts getting smaller, it is bulging on the other side.

Sharing observations

What the students have seen

- As the full moon keeps getting smaller, you can see a half-circle moon again in the sky.

- You can see this half-circle moon in the morning on the way to school.

- Soon, this half-circle moon looks like a crescent again.

- Sometimes you can only see the moon during the daytime.

Modeling the Moon Phases

In Ms. Cairo's class the moon-phase journals are not the end of the project. The next step is for the students to construct a model of the moon's phases. They use a volleyball to represent the moon, and they darken the room before they start.

Ellen stands in the center of a large space, representing the planet Earth. Patrick, holding the volleyball, walks around Ellen, keeping the same side of the volleyball facing her at all times. Johanna, representing the sun, stands to one side of the moon/Earth system with a strong flashlight, which she shines at the volleyball. As the volleyball reflects the light from the flashlight, these science ideas emerge:

Moving from observation to model

- You can light only half the volleyball (moon) at one time because it is a sphere.

Emerging ideas

- As it moves around the Earth, only part of that lit half is visible from the Earth.

- The moon rotates once as it revolves once around the Earth.

- The same side of the moon is always facing the Earth.

- Different parts of that side of the moon receive light at different times in its journey.

- The shapes we see in the sky are called *phases.*

- When we see the full moon, the entire side of the moon that faces the Earth is reflecting light.

By this time, the moon-phase journal and the following activity have led the students to formulate new questions. They wonder, for instance, about the yellow/orange color of a rising or setting moon; about the "halo" that sometimes appears around the moon; and why the moon appears bigger when it is low in the sky. These are complex issues, and the students explore them in resource books and on the Internet.

New questions

EXPANDING   MEANINGS

. . . .

The Teaching Ideas Behind This Story

■ We can learn a lot about nature simply by observing it.

■ Giving students more than a month for the project allows them to observe the moon phases' repeating themselves. Besides leading to better
understanding of the moon, this process helps the students give meaning to the notion of cycles in other natural processes and in their own
lives.

■ Although Ms. Cairo hopes the students will understand that cycles of
the moon's phases repeat and are predictable, she does not give them
this information in advance.

■ The information the students gain from keeping a moon-phase journal
makes it possible to construct the volleyball model of moon phases in
the classroom. Thus, like Mr. Johnston's edible solar system, the journal
is not just an end in itself, but also the basis for further explorations.

The Science Ideas Behind This Story

■ Like the planets, the moon is visible because it reflects the sun's light. In
other words, the light from the sun radiating outward in space in all directions bounces off the surface of the moon.

■ The moon appears to have different shapes in the sky as it revolves
around the Earth. These different shapes are called **phases of the moon**
(see Figure 10.3).

■ It takes 29 days for the moon to progress from one full moon to the next.
This is the time it takes the moon to make one revolution of the Earth,
and it is the same as the time it takes the moon to spin (rotate) once on
its axis. Because the moon rotates once as it revolves, we see only one
side of the moon from Earth.

■ As the moon moves around the Earth, different portions of its lighted
side are visible on Earth because of the moon's position relative to the
sun and the Earth.

FIGURE 10.3 **The phases of the moon.** Beginning with a new moon, when the side of the moon facing the Earth is entirely dark and invisible, the moon waxes (appears to grow) until full moon, when we see it as a fully lit circle. Then the moon wanes (appears to diminish), passing through phases that are mirror images of the earlier ones.

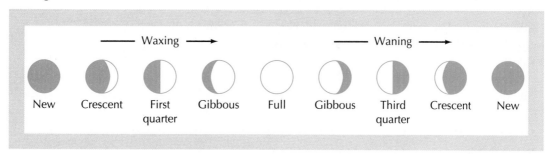

- As the moon appears to get bigger in its progress toward a full moon, we say it is *waxing*. As it appears to get smaller after the full moon, we say it is *waning*.

Questions for Further Exploration

- What other types of objects could model moon phases in the classroom?

- What do you think is meant by "Earthshine"?

- Have you ever looked at the night sky through strong binoculars or a telescope? What kinds of details do you think you would see that are not observable with the naked eye?

Resources for Further Exploration

Electronic Resources

SKY Online. **http://www.skypub.com**. This online astronomy site has current data on eclipses, comets, and other celestial events. Real-time photos and timely news clips are included.

Stone, Wes. *Skytour.* **http://www.lclark.edu/~wstone/skytour/skytour. html**. A wonderful site with a page devoted to moon phases, including rich explanations of "Earthshine," eclipses, and more. The site also has a section on planets, exploring Mercury, Venus, Mars, Jupiter, and Saturn and their visibility in the night sky.

Print Resources

Sneider, C. (1986). *Earth, Moon and Stars. GEMS.* Berkeley, CA: Lawrence Hall of Science. This activity book was developed for grades 5 through 9 to explore astronomy models and simulations, including moon phases and eclipses.

USING MOON-PHASE JOURNALS

Keeping a moon-phase journal has many similarities with keeping a science journal. As we noticed in Part One, journaling in elementary science invites students to observe nature in some way and then reflect on their observations. The moon-phase journal is most appropriate for students in fifth grade or higher, since this exercise requires that they be outside frequently at night. Although on some nights they may be able to see the moon through a window, it is important to alert families to this project and solicit their support and cooperation.

Recording both successful and unsuccessful experiences

During some parts of the month, the moon may not appear until after the students' bedtimes. As well, weather can interfere with viewing. To keep the students from feeling frustrated, it helps to ask them, as Ms. Cairo did, to record all their experiences, even the unsuccessful ones. And because the moon's period of revolution is about four weeks, it is important to define the time period as longer than four weeks; five or six weeks is a good idea.

It is my experience that students of all ages, from fifth grade up, take pride in their moon-phase journals. I have even used this project in my teacher education classes, where the preservice teachers become equally enthusiastic about the project. I explain that I hope the experience will enhance their confidence in their abilities to observe. I also make a point of giving them the freedom to express and present their observations in any

THINK OF THE TASK AS AN ADVENTURE, I TELL MY STUDENTS.

way they choose. "Think of the task as an adventure," I tell my students. "What can you find out about the moon simply by observing it on a regular basis?" Many are astonished to realize that just by looking closely, they can discover a great deal about a natural object they have seen all their lives.

The Smudge on the Moon

In the fall of 1996, during a six-week period when my teacher education students were observing the moon and recording their data in moon journals, I wrote the following in my own journal:

It was shortly after 9:00 P.M. on a cool autumn night. A deep purplish smudge seemed to creep upon the lower left edge of the full moon. The air was humid, and the moon shone brightly despite approaching clouds. Slowly, the curved smudge began to cover more of the moon's surface. I went out every ten minutes and looked toward the eastern sky—closer to overhead than to the horizon. By 10:30 P.M. the purplish smudge had covered the full face of the moon.

Eager to see what would happen when I glanced again, I went out at 10:40, and to my surprise the moon and its smudge were gone! The clouds were dense; the thick smoky-gray cover offered no peek of this bright moon emerging from the last lunar eclipse of the twentieth century.

Still, how exciting that this evening, right in the midst of our class's moon-phase journal project, there was a lunar eclipse! I hope everyone in the class was watching.

Sharing journals

When students bring their moon-phase journals to class, I find it useful to have them share the journals with each other before recording their observations. As they review and discuss all the information they have gathered after weeks of keeping a journal, the science ideas really begin to emerge.

Revealing hidden talents

If you start a moon journaling project with your students, you'll also find it has one other advantage. It will allow you to learn about some of your students' hidden talents. The artists and the poets in particular will have an opportunity to show their creativity.

Excerpts from the Moon-Phase Journals of Preservice Teachers

It's cool how the moon is always there. Even when it's cloudy or raining and we can't see it from the Earth, it's still up there! Kids need to understand that. As a kid, I just thought it disappeared!!

—Jeannie M.

Today was a clear, cool day. At 7:00 P.M., I got a wonderful view of the moon. You could actually see the circle shape of a full moon, but only a small portion of the moon filled it. (Like a silhouette.)

—Gail S.

A beautiful, mild evening. The sky is clear—a dark blue glow. The moon is bright. It's around 8:10 P.M. and the moon has "grown" since I last saw it. Almost half moon. The right edge is defined, while the left side seems wavy, unclear, and transparent.

—Ayanna B.

When I look at the moon long and hard I can see what appear to be land forms, maybe mountains. The moon looks like Earth does from a distance [in photographs taken from space]. Maybe those shadows or seeming land forms don't really exist on the moon? I'll have to look into this.

—Bess B.

A bright, clear blue—not a star to be seen.
A gibbous moon stands alone and gleams.
This odd shaped moon seems extra white,
With a frizzy ring around it of gentle light.

—Kimberley S.

I don't think I ever really knew what caused the "phases" until now, or why the moon seems to appear at different points in the sky. Actually, I never really thought about it until now. This assignment has certainly taught me to stop and look at things around me. I don't think that I will ever look at the moon in quite the same way again.

—Meryl F.

Where does the moon appear next? I think I know—I'm onto its pattern!

—Serena S.

All excerpts reprinted by permission.

EXTENDING UNDERSTANDING WITH MODELS ....

Thinking back to your own experience in science lessons, what did you do when the science topic did not lend itself to direct exploration? You probably remember using library research to accumulate your information. Although research of this type is useful, we know that direct observation and manipulation enhance the possibilities for students to construct their own meanings and extend their understanding.

Engaging students through models

As you have seen in this chapter, making models is an excellent way to engage students in observing and manipulating even when the ultimate subjects of study are not readily accessible. By manipulating the model, students can simulate operations in nature. Think of model building as a step toward model manipulation, which itself leads to further research, more model building, and deeper understanding.

Key Terms

model *(p. 210)*
solar system *(p. 218)*
planet *(p. 218)*
asteroid *(p. 218)*
star *(p. 218)*
orbit *(p. 218)*
ellipse *(p. 221)*
focus *(plural* foci) *(p. 221)*
satellite *(p. 223)*
perihelion *(p. 223)*
aphelion *(p. 223)*
year *(p. 223)*
phases of the moon *(p. 228)*

11 Expanding the Science "Box": Explorations of Electricity

Energy is a fundamental concept in the study of science. Often people use the word *energy* in daily conversation when they refer to not having enough energy or turning off the light to save energy. Yet, in spite of this ready connection to everyday life, energy remains an abstract concept in elementary school science because it is not a concrete substance with easily measurable properties, like matter is.

Energy exists in many forms, and comprehending that one form of energy may be transformed into another requires a high level of understanding. Elementary school students typically explore four forms of energy: sound, light, heat, and electricity. Their understanding deepens gradually with increased exposure, until by secondary school they can deal with fairly abstract concepts of energy measurement and transformation.

Since energy is such a complicated science idea, you probably think that many students develop alternative conceptions about it. You're right, as you'll see when you read the stories that follow. But this chapter also focuses on another type of alternative conception—one that exists in science education itself. It is found among teachers, administrators, creators of curriculum materials, and many others.

In the following stories, we examine the ways in which two fifth-grade classes explore the beginning of a unit on electricity. Both teachers have been given a typical science "box," that is, a kit of materials to use with their students. The kit has arrived complete with specific instructions for the teacher and little booklets for the students. As you read, think about the nature of scientific activity in these two classes. Look for moments when true inquiry is taking place. Ask yourself what kind of alternative conception about *education* is evident in one or both of these classrooms. And what does the chapter title mean by "expanding the science 'box'"?

| S C I E N C E   S T O R Y |

Batteries, Bulbs, and Wires

Let's visit Ms. Stone's fifth-grade classroom in a midwestern suburb. The students are about to begin a unit on electricity, and Ms. Stone is excited. The school district has just received several electricity kits from a commercial manufacturer. Ms. Stone is pleased that she will have these new materials to teach the unit on electricity.

The night before the unit begins, she reads the "Instructions to the Teacher" and prepares her lesson. When she examines the contents of the science kit, she finds packages of D-cell batteries, battery holders, small 1.5-volt bulbs that look like flashlight bulbs, holders for the bulbs, and spools of thin wire labeled "bell wire." The kit also includes several wire strippers and switches. Ms. Stone, following the directions in the kit, cuts 30-centimeter strips of bell wire and uses the wire strippers to remove the plastic insulation from the ends of each strip. With small self-sticking letters, she then labels one end of each wire strip "A" and the opposite end "B."

Ms. Stone's careful preparation

The next morning she explains to the students that they are going to begin the electricity unit and need some definitions. She has arrived early to write three definitions on the blackboard, which she now asks the students to copy into their notebooks:

Ms. Stone's definitions

An *electric circuit* is a continuous pathway for an electric charge or current to follow.

A *series circuit* is a simple circuit where the flow of electricity has only one path.

A *parallel circuit* is a circuit where the flow of electricity has more than one path from the same power source.

Ms. Stone reads the definitions aloud and goes over the words. Quietly the students copy these terms in their notebooks.

Ms. Stone's thinking: Ms. Stone believes that the students need the vocabulary in order to carry out the investigation meaningfully. She believes in giving the meanings up front, before the students begin the exploration.

Ms. Stone asks the students to divide into groups of four. Each group, she explains, will get four bulbs, two batteries, four bulb holders, and two battery holders. Also each group will receive seven stripped wires, each 30 centimeters long, with the ends labeled "A" and "B." One end of each battery holder is labeled "B," and the other end is labeled "A." She explains to the students that they are *not* to touch the materials until she gives them instructions.

When all the materials are distributed, Ms. Stone begins. "Now, will one person in the group take the batteries and place them in the battery holders? Next, another person should gently screw each bulb into the bulb holders."

The students do as they are instructed. Then Ms. Stone continues, "Does everyone see that one end of the battery holder is marked A and one end of the wire is marked A? Okay, take turns as we follow the next steps. Attach the end marked A on the wire to the end of the battery holder marked A by putting the wire through the loop and twisting to make good contact. Now repeat what you just did with another wire, only this time take the end marked B and loop it through the end of the battery holder marked B.

"Now you should have one battery that has two wires coming from it, one from each end," says Ms. Stone as she holds up a sample of what she means. The students are fidgety, especially those who are not working directly with the materials at that moment.

"The next thing we are going to do," Ms. Stone continues, "is to connect the end of the wire that has a B on it to one side of one of the bulb holders. . . . Then take another wire and connect the end marked A to the other end of that bulb holder. . . . Now take the end of that wire marked B and connect it to a second bulb holder. . . . Then take the A end of the other wire that is attached to the battery, and connect it to the other side of the second bulb holder."

Ms. Stone talks slowly and waits for the students to complete each step before proceeding with the next step. Finally, the bulbs light!

Ms. Stone's thinking: Ms. Stone is pleased that the students have followed the directions well. She notices that every group's bulbs worked.

"Okay," says Ms. Stone. "You have made a simple series circuit. Let's look at the definition on the board." The students listen as Ms. Stone reads

Do you agree with Ms. Stone's thinking?

Each group's materials

Thorough instructions

Why are the students restless?

Success?

the definition again. They are instructed to draw their series circuit and label it in their notebooks.

After Ms. Stone checks their drawings, she continues, "Now we are going to make a parallel circuit. This is a little more difficult, so you have to listen carefully." She instructs them to keep their series circuit intact. "Take the second battery and connect the remaining wires, matching A ends of each wire and B ends of each wire to the ends of the battery holders." When the students have done this, she notes, "Now each end of the battery holder should have two wires coming out of it." Ms. Stone checks that each group has connected the wires correctly.

"The next step," she goes on, "is to take one free bulb in its bulb holder and attach end B of one wire to one side of the bulb holder and end A of the wire on the other side of the battery to the other end of the bulb holder. You should all have one lit bulb now. Next, take the last bulb and connect it to the open ends of the remaining wire as you just did."

The two bulbs light. Once more, all the groups get the same result.

"This is called a parallel circuit," Ms. Stone explains. Together, she and the class read over the definition of a parallel circuit.

"Tomorrow," Ms. Stone tells the students, "we will use our series circuit to test for materials that carry electricity."

New circuit, similar procedure

WHETHER WORKING WITH MANDATED CONTENT AND ACTIVITIES . . . OR CREATING ORIGINAL ACTIVITIES, TEACHERS PLAN TO MEET THE PARTICULAR INTERESTS, KNOWLEDGE, AND SKILLS OF THEIR STUDENTS AND BUILD ON THEIR QUESTIONS AND IDEAS.

—National Science Education Standards

SCIENCE STORY

Batteries, Bulbs, and Wires Revisited

Ms. Travis is a fifth-grade teacher in another school in the same school district. She is working with the same materials as Ms. Stone—with one difference. She has requested funds from the fifth-grade science budget to purchase 12 flashlights and 24 C batteries, two per flashlight. She assembles all the flashlights with their batteries. When the students enter the room the following morning, she invites them to form science groups of four.

Before starting the activity, however, Ms. Travis asks the students what they would purchase if she asked them to go to the supermarket and pick up a pound of electricity. The students call out different answers. One student says she would buy a pound of light bulbs. Another says he would buy a pound of batteries. Ms. Travis listens and asks them why they made

Extra materials

A pound of electricity?

those choices. "Is it the same to say that light bulbs *are* electricity," she wonders aloud, "as it is to say that light bulbs *work on* electricity?"

Ms. Travis's thinking: Ms. Travis knows that students often think of the materials that produce or carry electricity as the electricity itself. This alternative conception is common because electricity, a form of energy, does not have mass or volume like matter. Ms. Travis is trying to bring out prior ideas of this sort as the students begin the electricity unit.

Bringing out prior ideas

The students listen carefully to Ms. Travis's question. One student remarks that you can't really buy a pound of electricity because it isn't a real thing. Ms. Travis prompts, "Can you say more about that?"

"Well," the student continues, "you can't really hold electricity in your hand, but it can give you a shock." Ms. Travis invites others to think about that.

ENERGY IS MORE DIFFICULT TO GRASP THAN MATTER BECAUSE WE CANNOT HOLD IT IN OUR HANDS.
—*Janice Koch*

She then shares the following idea: "We know that anything that has mass and takes up space is matter. So if electricity is not matter, what is it?" This class knows she means "energy" because they have spent weeks studying heat energy. But it is important to Ms. Travis to remind the students again that energy is more difficult to grasp than matter because we cannot hold it in our hands.

Now they are ready to begin the activity. She explains that she would like them to take apart some flashlights. "Take these apart; take out the inside parts," she requests as she distributes two flashlights to each group of four students. "It is better to work with one partner for your exploration just now," she adds. "How many of you have opened and explored flashlights before?" About half the students raise their hands.

Beginning an activity

"Girls and boys, you will notice that the top unscrews and then several pieces come out, including the bulb. Sometimes those pieces fall out, so be careful not to hold the flashlight too high. As you and your partner examine the insides of the flashlight, draw and label the parts in your science journal."

The students take apart the flashlights and make rough drawings in their journals. Then Ms. Travis distributes strips of bell wire, one to each pair of students. "Let's see if you can use this wire, with one of the batteries and the bulb you have found, to get the bulb to light. Before you begin, let's look at these tools."

A challenge to the students

She holds up the wire strippers. "Where have you seen these before?" Some students recognize the wire strippers from tool kits they have at home. "Why are they important for our experiment?" Ms. Travis asks. Some students, who have prior knowledge about stripping wire, offer answers. Ms. Travis distributes the wire strippers and suggests that the students use them to bare the ends of their wires.

Ms. Travis's thinking: *Ms. Travis wants her students to learn by doing. How-ever, she knows that the battery, bulb, and strip of wire will be cumbersome to ma-nipulate, and she doesn't want the students to get so frustrated by lack of success that they lose interest. Thus she suggests that they strip the ends of their wire, giv-ing them a better opportunity to solve the problem.*

Suggestion—but no instructions?

The groups of students make several attempts before finding the ways that work. Most groups discover that the bulb will light if one end of the wire is wrapped around the bottom of the bulb and touching one end of the battery, while the other end of the wire is touching the other end of the battery. One group of students succeeds by stripping a portion of the cen-ter of the wire, wrapping that portion around the tiny bulb, and then touching either end of the battery with the free ends of the wire.

Some efforts succeed; others fail.

Students visit each other and try several ways to get the bulb to light. Ms. Travis asks them to draw the ways that work *and* the ways that do not work.

She then proposes that the students use *both* flashlight batteries and the wire to light the bulb. Manipulating the two batteries and one wire is tricky. But when they finally manage to get the bulb to light, they notice it is much brighter than it was with only one battery.

Another proposal

These students are conducting their own investigation with bat-teries, bulbs, and wires. They are in-vested in solving their particular problem.

Ellen Senisi/The Image Works

Ms. Travis asks the students to reassemble their flashlights, turn them on, and compare the light from the flashlight with the light that they have just created with the two batteries outside the flashlight. The students notice that the brightness seems to be identical.

For homework, she asks them to write what they have done this morning in their science journals and to list some "rules" they have discovered for getting their bulb to light, first with one and then with two batteries.

Journal writing

The next morning, Ms. Travis empties the contents of the science electricity kit on the front table of the classroom. "Let's look at the materials we have to work with here. Can we identify some of them?"

Opening the kit at last

The students recognize the bulbs as looking just like the bulbs in the flashlights. The wire, too, is identical to the wire they used the day before. Although the batteries are somewhat larger, the students have no trouble identifying them as batteries. But this kit also has battery holders and little bulb holders. Some students have seen these before, and others have not.

Some objects familiar, others not

Ms. Travis says, "Each science group should take two bulbs, two bulb holders, a battery, a battery holder, and 60 centimeters of bell wire to start. You will also need a wire stripper. Our class's problem is this: How can we get both bulbs to light at the same time with as much brightness as possible? Start to explore your materials carefully. Notice how the battery holders and the bulb holders are constructed. Brainstorm with one another about where to attach your wires. Use the rules you developed for homework last night. Then make a plan with all the members of your group. You will want to try out lots of ideas. Take your time!"

A problem to solve

Ms. Travis's thinking: The students are excited about handling the materials, but Ms. Travis knows that connecting the wires to the bulb holders and the battery holders is sometimes tedious. Planning their procedure beforehand may save the students unnecessary labor. Also the discussions among themselves should help them learn.

The students gather their materials and develop their plans. Some have a sense of where to go; some ask Ms. Travis where to begin. She offers responses such as, "You may want to screw the bulbs into the bulb holders."

The students make plans.

The students start stripping the ends of the wires. The groups are busy and engaged. Walking around the room, Ms. Travis notices varying levels of comfort with the materials. She points things out and makes further suggestions. She reminds one group, "You may have to cut that strip of wire. You may need more than one wire." To the class she announces that they can have more wire if they need it.

Additional suggestions

Ms. Travis's thinking: In walking around the room, Ms. Travis wants to see how the students are thinking. She is hoping that their experience the previous day will help them construct their own electrical connections with these new materials.

Using three pieces of wire and connecting the first wire from the battery holder to a bulb, the second wire from the first bulb to the second bulb, and the third wire from the second bulb back to the battery, most of the groups get the bulbs to light. They notice that the lights are very dim—much dimmer than when they used one bulb and one battery the day before.

The students' solutions

After applauding their efforts, Ms. Travis begins to provide some labels for what they have created. "Can we label the pathways we have made with these batteries, bulbs, and wires?" she asks. "What do scientists call this type of path?" Some students shout out "circuit," and Ms. Travis writes the word on the chalkboard. Together, she and the students define a circuit as a complete pathway from one end of the battery through the wire and the bulbs to the other end of the battery and through the battery again.

Collaborating on a label and a definition

This pathway, she tells them, is a continuous route for tiny charged particles called electrons. "These electrons carry the electric charge through the circuit. All these charged particles flowing along the wire is what is meant by an electric current." On the board she draws a sketch of the circuit that the students have constructed (see Figure 11.1).

At this point she slips in a new question: "I hope this helps to explain why our bell wire is *coated*. Let's think about that as we work."

A new question—but no answer

Ms. Travis's thinking: At this point, Ms. Travis wants the students to associate the correct labels with the pathways they have created. She is not interested in

FIGURE 11.1
. . . .

A series circuit with a battery in a battery holder, two bulbs in bulb holders, and three wires.

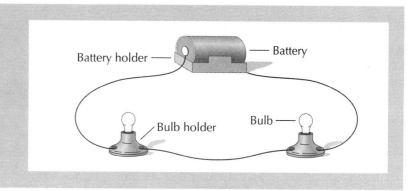

brainstorming about the coating on the bell wire at this time; she just wants to plant the seed of this idea, to see if it will germinate as they work with the wire.

Now Ms. Travis challenges the students to create a circuit that could enable the bulbs to be brighter while both are still lit at the same time. She invites them to take more wire if needed. Once again, she circulates among the groups, directing and coaching, probing the students' thinking and encouraging them when they get frustrated. "You may want to try one bulb at a time," she says to one struggling group. To another group she remarks, "Draw what you have done here. It really seems to work." With yet another group, she invites them to visit a group in the back of the room that has been more successful. All this time, students are trying to figure out how to light two bulbs at once with a single battery without making the bulbs so dim.

A further challenge

After a while, Ms. Travis asks one member of each group to come to the board and draw the circuits they have constructed. The solutions that work all involve using four wires, connecting each bulb to the battery independently of the other (see Figure 11.2). The difference in brightness is very noticeable. The two bulbs connected independently of each other are much brighter than the two bulbs that shared the same circuit.

Sharing solutions

"It is time to label the two types of circuits you have made," Ms. Travis tells the class. "In this one"—she points to the drawing she made of their first circuit (resembling Figure 11.1)—"the electric charges travel in a single path, through the battery, the wires, and both bulbs. This is called a *series circuit.* In the second type of circuit that you made"—now she points to their own drawings (resembling Figure 11.2)—"each bulb is connected

Adding labels—*after* the circuits are built

FIGURE 11.2

A parallel circuit with a battery, two bulbs, and four wires.

Because the bulbs are connected to the battery independently of one another, they will shine brighter than the bulbs in Figure 11.1.

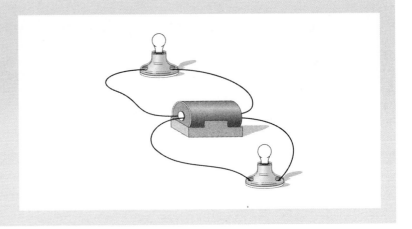

in its own separate pathway. This is called a *parallel circuit*. Now, let's take some more batteries and bulbs and wires and set up both of these types at the same time."

The students set up a series and a parallel circuit, with one battery and two bulbs in each. Ms. Travis asks them to find different ways of extinguishing one or both of the bulbs. She asks the students to write about their findings in their journals.

After some exploration, the students learn that, in the series circuit, any break in the pathway causes both bulbs to go out. In the parallel circuit, though, each bulb is affected independently, and for both bulbs to go out, there must be a break in each of the separate pathways. Further, in the series circuit, the children notice that when they unscrew one bulb, the other one goes out. In the parallel circuit, when they unscrew one bulb, the other bulb stays on.

Varying the experiment

Before the morning is over, Ms. Travis asks the students in each group to come up with further questions that they want to investigate. "What are you most dying to find out? List the questions that your experiments with batteries, bulbs, and wires can help you figure out. Make sure that you also list the materials you will need."

Here are some of the students' questions:

The students' own questions

Are the bulbs always brighter in a parallel circuit?

What happens with two batteries in a series circuit? In a parallel circuit?

What happens to the brightness in a series circuit with three bulbs?

What is a light bulb made of?

What happens when we combine circuits with other groups?

Ms. Travis lists the questions and the name of the child posing it on poster paper in the front of the classroom. The children will be exploring their own questions for many days.

The Next Day: Lighting a Shoebox House

The following day Ms. Travis brings in a large shoebox that is set up like a little dollhouse room. It has cardboard chairs and a little sofa. On one inside wall is a picture of a fireplace. Through a hole in the top of the box, a miniature bulb (like the flashlight bulbs the students have been using) hangs as a ceiling light. On the outside of the box, the bulb holder is connected to two long wires attached to a battery. The light works! Unlike a normal household light, however, this light has no switch, and the students are given time to think about what a switch might look like.

A simple household lighting model

Plugging In

Peg plugged in her 'lectric toothbrush,

Mitch plugged in his steel guitar,

Rick plugged in his CD player,

Liz plugged in her VCR.

Mom plugged in her 'lectric blanket,

Pop plugged in the TV fights,

I plugged in my blower-dryer—

Hey! Who turned out all the lights?

—Shel Silverstein

Source: Copyright © 1996 by Shel Silverstein.

The children love the shoebox room. Ms. Travis now brings out several switches for their circuits, and she invites each group to experiment with circuits and switches. (You may remember that switches were included in the kit of materials that all the fifth-grade teachers in this district received.) How, Ms. Travis asks the students, could they make her shoebox room brighter? They all suggest another bulb, but the question of how to connect the other bulb becomes a point of discussion.

A puzzle based on the model

Ms. Travis demonstrates the effects of two bulbs in a parallel circuit lighting the room as compared with the same bulbs in a series circuit (see Figure 11.3). The students immediately notice that the parallel circuit makes the room brighter, just as their two bulbs were brighter in a parallel circuit the day before.

As the class's electricity investigations continue, Ms. Travis offers a list of questions for the students to answer at home:

Questions to explore at home

Is your home wired with fuses or with circuit breakers?

Can you ask a parent or guardian to show you the fuses or circuit breakers?

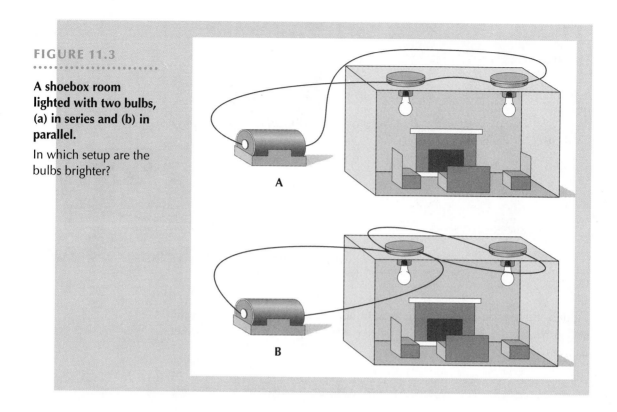

FIGURE 11.3

. . . .

A shoebox room lighted with two bulbs, (a) in series and (b) in parallel.

In which setup are the bulbs brighter?

A

B

What types of things in your home work on electricity?

Do you have an idea about which appliances may be connected in a series circuit in your house and which ones in a parallel circuit? How could you find out?

In addition to answering these questions at home, the students pursue many interesting activities in the classroom during the following days. Although the groups do not all do the same things at the same times, each group reports its discoveries to the class at large. For example, one group discovers that the more parallel circuits they place on a battery, the faster the battery will burn out. Another group uses a series circuit to make a tester, and they test which materials conduct electricity (see Figure 11.4). Two groups create a series circuit around part of the room. It has four batteries and eight bulbs, and the students demonstrate how all eight bulbs go out when a single bulb is unscrewed. Ms. Travis also engages the students in building things with their circuits. Some students, following the model of a shoebox room, construct a

Groups pursue
different investigations

FIGURE 11.4

A simple electrical tester.

lighted three-room shoebox house. Others make a model flashlight; still others build a blinking lighthouse and a quiz board with secret circuits.

EXPANDING MEANINGS

The Teaching Ideas Behind These Stories

- Notice how Ms. Stone gives information at the beginning of the lesson, while Ms. Travis waits until the students have developed some understanding before providing new language.

- Ms. Stone engages students in the exploration of electrical circuits on *her* terms. She controls the activities, the time students spend on each step, and the outcomes they should achieve. Ms. Travis controls much of the experimental design and the initial problem posing, but then gives the students control over procedures and findings.

- Ms. Stone is concerned that everyone "get" all the science ideas from her, at the same time. Ms. Travis is certain that all the science ideas will emerge from the students, pieces at a time. She relies on reporting from student groups to address ideas that do not emerge from the initial activities.

■ Ms. Stone's procedures are lockstep and rigid. There is no room for students' own ideas about how to explore. Ms. Travis, in contrast, allows a diversity of procedures. She thinks this difference is important for learning.

■ Ms. Stone does not provide personal connections or a personal context for the electricity unit. Ms. Travis provides such contexts by engaging the students with flashlights, a mock dollhouse room, and questions about their home wiring.

■ All in all, Ms. Travis's students are likely to learn better and more deeply because they are conducting their own investigations and becoming invested in answering their own questions.

The Science Ideas Behind These Stories

■ **Electricity** is a form of energy. In some respects scientists are still unsure of exactly what electricity is, but the word *electricity* is used to describe a flow of electrons.

■ **Electrons** are negatively charged particles that are found in the atoms of all elements. Each tiny electron is thought to be a particle of negative electricity.

■ An **electric current** is a flow or motion of electrons. Electric currents in wires are caused by electrons moving along the wire.

■ An **electric circuit** is a pathway for a current. A circuit requires some source of electrical power, such as a battery or generator; a pathway for the electrons, such as copper wire; and some appliance that uses electricity, like a light bulb or a bell.

■ In a simple electric circuit using a light bulb and a battery, the electrons flow from the negative (flat) end of the battery through the wire to the bottom of the bulb (or the side of a bulb holder), through the filament inside the bulb, and out the bottom of the bulb (or the other side of the bulb holder), through the other piece of wire, to the positive (bumpy) end of the battery. The current then passes through the interior of the battery to complete the circuit. There is a continuous flow of energy

through these devices unless some part of the circuit is stopped or broken—for instance, if a bulb burns out, a wire gets loose, or a battery is
used up.

- In a **series circuit**—the type of simple circuit just described—the electric
current has only a single pathway through the circuit. This means that
the current passing through each electrical device is the same. Each time
an electrical device, like a bulb, is added to the series circuit, the total
amount of electrical force must be shared among a greater number of
devices. That is why each bulb becomes dimmer when a new bulb is
added.

- In a **parallel circuit,** in contrast, each device has a separate pathway—
its own separate branch of the circuit. Thus, if one bulb goes out, the
other bulbs or devices in a parallel circuit are unaffected.

- Each device in a circuit offers a certain amount of *resistance* to the current. In a series circuit, two bulbs present twice the resistance of one
bulb. Since the current equals the force divided by the resistance, each
bulb gets only half as much current. In a parallel circuit, on the other
hand, there is only one bulb per path; therefore, there is less resistance
on each path, the current supplied to each bulb is greater, and the bulbs
glow more brightly. This also means that the battery will burn out twice
as fast, because it is supplying twice as much current as in the series circuit.

Questions for Further Exploration

- Explore a small flashlight bulb to see how it lights.

- Why does it stand to reason that homes are wired in parallel and not in
series circuits?

- In the experiments described in our science stories, why is it important
to use bell wire, which is copper wire coated with a plastic sheath?

Resources for Further Exploration

Electronic Resources

Bill Nye the Science Guy. **http://www.Disney.com/DisneyTelevision/
BillNye**. Follow the links to Episode 18 in physical science for an electrical
experiment geared toward simple concepts in electricity.

One Hundred Years of the Electron. **http://www.brakenhale.berks.sch.uk/ projects/science/electron/title.htm**. This site, from the Brakenhale School in the United Kingdom, has an electricity link that provides an excellent exploration of electricity and the history of its discovery and applications.

Print Resources

Markle, S. (1989). *Power Up: Experiments, Puzzles and Games Exploring Electricity.* New York: Atheneum. A book full of neat ideas to enhance any electricity unit.

National Science Resources Center. (1991). *Electric Circuits STC.* Burlington, NC: Carolina Biological Supply Co. Designed for grades 4 and 5, this series of activities explores electricity and circuits. The projects include wiring a cardboard house to light each room. The teacher's guide is useful; the student activity books are not.

THINKING ABOUT TEACHING AND LEARNING

Ms. Travis's belief system

Let's think further about the two teachers whose classrooms we have visited. Ms. Travis holds out the expectation that her students are knowers, that they have their own ideas and are capable of constructing and carrying out investigations on their own. This is Ms. Travis's belief system, and it influences her teaching practice. Her teaching is a good example of the principles that this book advocates. She expands the science "box"—in both a literal sense, by letting her students use more materials and explore a wider range of activities than the school-supplied kit allowed, and in a metaphorical sense, by releasing herself and her students from the intellectual confinement of prepackaged instructions.

Ms. Stone's belief system

But what about Ms. Stone? Clearly, Ms. Stone works hard, has respect for her students, and cares deeply about their instruction. She is, however, unwilling to trust that her students can come up on their own with the activities that are neatly laid out, step by step, in the student workbooks in the fifth-grade electricity kit. In Ms. Stone's personal model of teaching, she is the authority in the classroom. She believes that this is what the children require. She believes that they look to her for the right answers and that she is not doing her job unless she provides these answers.

This attitude is not uncommon, and it is not at all ill intentioned. Well-meaning elementary school teachers all over the country believe that students learn only if the teacher tells them what and how to learn.

DOES A TEACHER WHO GUIDES AND MEDIATES DO *LESS* THAN A TEACHER WHO "DELIVERS" INSTRUCTION?

—*Janice Koch*

Ms. Stone's type of authority

You may be wondering, "Well, what is teaching if we don't tell them what and how to learn? What is left for us to do?" Think back to Chapter 4's discussion of the teacher as mediator, and consider whether Ms. Travis is really doing *less* than Ms. Stone. Remember all the other science stories in this book that have shown teachers working very hard with their students. I think you'll find that the teacher who guides and mediates learning is doing lots of planning and active listening in ways that are not required of a teacher who merely "delivers" instruction. Also, even though Ms. Travis does not behave like an absolute science authority in the classroom, she is actually taking on a great deal of authority because she herself assumes the responsibility of constructing the science experiences for her students. It requires a lot of self-reliance—as well as thoughtful consideration—to take a package or kit of materials, discard the prescribed instructions, and ask yourself, "How can I use this kit to create a meaningful learning experience for *my* students?"

As a teacher who guides and mediates, you are the person in the elementary school classroom who provides the most important conditions for real learning to occur. For example:

Your function as a teacher

- You provide a safe, caring environment.
- You help maintain classroom rules and procedures in predictable and comforting ways.
- You provide meaningful experiences and opportunities for problem solving.
- You encourage students to explore problems that have personal relevance for them as they go along. These problems become the basis on which students construct new understandings.
- You coach the students when they reach alternative conceptions that require further exploration.
- You provide information when appropriate and necessary.
- You hold out the expectation that your students are capable.
- You delight in their thinking processes and encourage them to have confidence in those processes.

Those are principles that apply to teaching any subject. When we look specifically at the teaching of science, we can extend the list:

Your role in teaching science

- As the teacher, you construct the ways in which you present the materials and invite the class to experiment with them.
- You delight in your own explorations of natural phenomena and embody that enjoyment for your students.

- Understanding that science ideas are complex, you help the students pull them apart rather than rush to oversimplify.

- You make connections as often as possible between the formal curriculum and the students' lives.

- You honor the gendered and multicultural ways in which your students see nature and natural events.

- You organize the science table, corner, or center in your room to reflect *your* students' interests.

MOVING AHEAD

No need to worry

At this point, rather than thinking that a teacher like Ms. Travis doesn't do enough, you may have the opposite reaction. You may read the above lists and think, "Wow, I can't do all *that*!" But you needn't worry. As we move ahead to Part Three of this book, we will explore particular ways to approach the task of creating constructivist learning environments for elementary school science. You will be comforted to learn that there are many tools to help you plan your lessons, conduct some self-evaluation, and assess your students' learning. Most important, there are specific steps you can take toward becoming an effective elementary science teacher.

Remember that as a teacher, you will be the facilitator of the learning process, responsible not just for opening the science "box" but for deciding how the materials and activities can best work for your students. Part Three gives you some guidelines for doing that.

Key Terms

electricity *(p. 247)*
electron *(p. 247)*
electric current *(p. 247)*
electric circuit *(p. 247)*
series circuit *(p. 248)*
parallel circuit *(p. 248)*

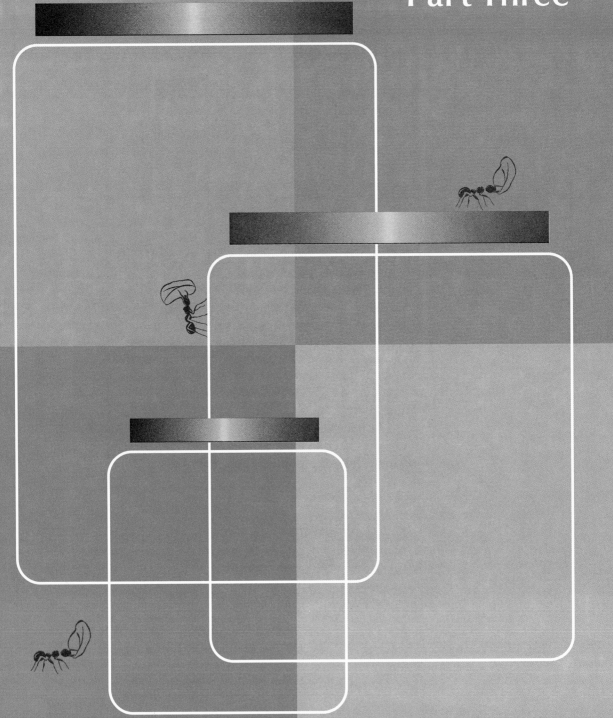

Part Three

Creating the Science Experience in Your Classroom

I n this final part, we will look at some practical matters that you need to think about as you prepare to teach science in elementary school. Part Two showed you what school science "looks like." Part Three demonstrates how to do it in your own classroom. It addresses lesson planning and questioning strategies, the use of technology in the classroom, selection of materials to match a curriculum, assessment, community involvement, and self-evaluation techniques.

Many of the points I make in the following chapters refer to the stories you read in Part Two. Since all of these issues will become very pressing as you begin your teaching career, you may want to refer to this part of the book often. As you gain experience, increasingly you will be able to use the ideas and suggestions in Part Three to create your own personal style in the classroom.

12

Planning for Science: Lesson Plans and Instructional Strategies

FOCUSING QUESTIONS

■ What is the difference between a science activity and a science lesson?

■ Do you prefer to learn independently or in small groups?

■ How do you plan for the unexpected?

A friend of mine used to say, as each school year began, "Remember, Janice, teachers can never be overprepared." As I gained experience in teaching, I discovered what he meant.

Being prepared means, first and foremost, having done your planning. The role of planning for elementary school science cannot be overstated. To develop a plan that guides your behavior and instructional practice, you need to give serious consideration to the learning experiences you want to provide, and then you need to organize and structure those experiences in ways that make sense to you.

Back in Chapter 2, I mentioned that we teach "who we are." That applies to planning as well as to your performance in class. When you come to school prepared, the students recognize that, and they develop an image of you as hard working and caring. This image helps you create a sense of community in the classroom.

This chapter looks at the plans you need to make for science lessons, as well as at instructional strategies closely related to your lesson planning. In Chapter 14, which discusses curriculum, we will explore larger-scale planning for an entire science unit.

AN ACTIVITY IS NOT A LESSON

Planning the activity

A major part of planning is setting up the activity your students will pursue. As the chapters in Part Two demonstrated, it is only by providing students with meaningful science activities—such as experimenting with liquids, planting seeds, or investigating snails—that we can lead them toward critical exploration of their world. Planning therefore involves acquiring and organizing the appropriate materials and deciding how the class will approach the activity.

Planning for ideas, skills, reflection

These steps, however, are just the beginning of the planning process. Although students may have a lot of fun experimenting with the materials you give them, *an activity is not a lesson.* You need to know what science ideas you expect your students to develop and which process skills (observing, predicting, measuring, and so on) they will be cultivating. Most important, it is not just the activity but the process of reflection that can lead to the construction of new ideas. Your responsibility is to facilitate the thought experience as well as the active experience.

The process of performing the activity and reflecting on it may be considered the science **lesson.** The type of lesson planning you do, as you saw in Chapter 11, will inevitably reflect your own ideas about how children learn. This chapter cannot give you a lockstep procedure for every occasion. But it can offer a general model for planning that can help you become a facilitator of your students' construction of meaning. In the sections that follow, we will begin with the lesson plan itself and then address the role of questioning, the use of cooperative learning groups, and some guidelines for reflecting on your teaching experience.

Planning to drop the plan

One more introductory point: Careful planning also includes the plan to drop the plan! This may sound like a contradiction, but when students are engaged in genuine inquiry, one question often leads to another. Your students' spontaneous curiosity will give you excellent opportunities to change your original plan and mediate their experience on *their own terms.*

PLANNING THE LESSON

Let's imagine that you are planning a new science lesson for your class. What type of **lesson plan**—that is, a document that describes your plans for the lesson—will help you engage your students in a meaningful science activity *and* invite them to think about, reflect on, and construct ideas from this activity?

There are many lesson plan templates—guides to help you get started. These are usually designed to predict the behavior of your students and of you as a teacher. But I have dedicated a great deal of this book to helping you understand that you *cannot* always predict where a meaningful science activity will lead. The goal, then, is to create a plan that allows for flexible procedures and critical thinking about the lesson.

How to plan for flexibility?

The guide that follows is a useful model and can get you started in the planning process. Over the years, it has been modified by several elementary science teachers as a result of their actual experiences in doing science with children. As you begin working with other teachers and see different ways of planning, you too can modify the plan to meet your own personal needs.

A Planning Guide

The box lays out the steps in a basic model for lesson planning. Each step is a question that you ask yourself. To explore how to use these steps, let's say that you are writing a lesson plan for a land snails experience like the one described in Chapter 8. The students will be exploring the snails' responses to the external stimuli of food, light, and water.

Goals: What am I hoping the children will get out of this science experience?   This part of your plan asks you to state the problem or phenomenon the students are exploring and what you are hoping they will discover. In the snails lesson, you might say:

Stating your goals

I am hoping that the students, through observation of their land snails, will notice that the snails respond to external stimuli and that the snails' antennae assist them in sensing their environment.

Science Activity: What will the students be doing on behalf of their own learning?   Here, you need to describe the activity briefly. For the snails lesson you might write:

Describing the activity

The students will work with three kinds of external stimuli—lettuce leaves, light, and water—to investigate the snails' responses to each stimulus. Each group will have magnifying lenses, lettuce leaves, water, and a light source. The students will test the three stimuli and record the results. The magnifying lenses will help them observe the snail's antennae.

A Guide for Making a Lesson Plan

Answer fully each of the questions listed below.

Goals	What am I hoping the children will get out of this science experience?
Science activity	What will the students be doing on behalf of their own learning?
Science ideas	What are the science ideas at the heart of this experience?
Process skills	What process skills will the students be using?
Materials	What materials do I need for the activity?
Making connections	How does the activity connect to what we have been doing in class? To the children's lives?
Teacher's role	What am I going to do to facilitate the science experience?
Pulling it together	How will I help the students organize their thinking and pull this lesson together?
Evaluation	How will I know what the students learned?

Stating the science ideas

Science Ideas: What are the science ideas at the heart of this experience?   In this step, you state the science concepts—the underlying ideas about the natural world that you are hoping the students will construct as they engage in the activity and in reflection about it. For example, in the snails lesson, you could write:

Snails are attracted to food and water, but tend to be repelled by light. Their behaviors are called "responses," and the lettuce, water, and light are called "stimuli."

Listing the process skills

Process Skills: What process skills will the students be using?   This can be a checklist for you, a reminder that science is both a method and a set of ideas. In the snails lesson, the skills are:

Observation

Inference

Recording data

Comparing and contrasting

You could structure the lesson to include other skills as well—for instance, you could add the process skill of predicting by asking the students to write down beforehand what they think the snails will do in response to each stimulus.

Materials: What materials do I need for the activity?   This section lists the materials that you need to organize before the lesson. In the snails lesson, the list is as follows:

Listing materials

garden snails	eyedroppers
lettuce leaves	water
magnifying lenses	flashlights

It is important, of course, to add quantities to your list so that you have enough materials for all the students to participate. How many flashlights and magnifying lenses? How much lettuce?

Making Connections: How does the activity connect to what we have been doing in class? To the children's lives?   Here you plan for real-life connections. To connect the snails lesson to the children's lives, you might ask the students about their own responses to food, water, and light. Or you might ask them how their household pets would react to these stimuli. Or you might invite them to think of other stimuli and their own responses.

Preparing to connect to students' lives

Teacher's Role: What am I going to do to facilitate the science experience?   This section addresses your role in the classroom. You may want to include key questions that you will ask the students during the lesson—for example, questions that will encourage them to use certain process skills. For the snails lesson, these questions could include:

Designing your questions

What do you notice about the snails when the table is wet with drops of water?

How can you tell what the snails like?

Later in this chapter we will look more closely at the types of questions you might want to ask.

As part of planning your role, you may also want to state that the students will be working in groups and you will visit each group. Any advance thinking about your methods as a guide and mediator belongs in this section of the lesson plan.

Teachers can never be over-prepared. Here, a teacher writes her plan for the students before they arrive to school. Some teachers prepare—daily or weekly—in the evening.

Mary Kate Denny/Photo Edit

Pulling It Together: How will I help the students organize their thinking and pull this lesson together?   This question refers to the closing part of the science lesson, when you give students the opportunity to formulate and express some of the science ideas they have developed from their experiences. For the snails lesson, you might write the following:

Have students talk about their results with the lettuce, the flashlights, and the drops of water.

Chart each group's responses on poster paper.

Arranging to pull together ideas

Evaluation: How will I know what the students learned?   Here you can plan an activity, a journal writing exercise, or a formal assessment. You will want to see how the students' learning fulfills your original goals and any further goals that have evolved along the way. To evaluate the snails lesson, for example, you could ask your students to write a story about their snails, including their observations of the snails' responses. This assessment will help you plan for future lessons.

Planning an assessment

If I Know It, Do I Have to Write It?

In the preceding section, you probably noticed the assumption that you would be *writing* your lesson plan. But you may have been thinking, "Is it really necessary to write the plan on paper or on my computer? Isn't thinking about it enough? After all, I'm not going to stop in the middle of the lesson and pull out a written plan."

Unfortunately, burdensome as it may sound, writing your lesson plan is important, whether or not a written plan is required by your school. It is a way to collect and represent your thoughts, then save and modify them as you go along. Writing the plan is another step toward becoming a reflective teacher. The document itself should be thought of as part of your own teaching journal. It becomes a tool to reflect on, use again, and change over time.

I still have lesson plans that I wrote the year I began teaching. The papers, now yellowing, remind me of how my thinking has changed in some ways and remained the same in others. I know that the process of writing these plans helped both me and my students, and it continues to do so.

Writing as part of reflection

Time Allotment

Your planning instrument becomes a tool for creating a science lesson in which the students have the opportunity to explore, reflect, and explore again. Sometimes one lesson plan takes several days or more to implement. As you saw in earlier chapters, especially Chapters 8 and 9, allowing time for sustained inquiry is vital to genuine science learning.

In the long run, your experience with your own teaching and with your groups of students will help you plan the allotment of time. As a general guide, though, here are two principles to follow:

Tips for allotting time

1. Allow at least 1 hour of class time for engaging students in any scientific activity, even a fairly simple one.

2. Always build in the flexibility to spend more time on an activity than you originally thought would be required.

After the Planning, the Letting Go!

Now that you have planned for the science lesson and organized the materials, how prepared are you to abandon your plan and respond to the

A Planning Checklist

Before you complete your lesson plan for the science experience in your classroom, the following checklist can help you decide if you are fully prepared.

☐ Have I done enough research of my own? Do I understand the science ideas behind this topic?

☐ Do I need special arrangements for live materials?

☐ Do I need parental or administrative permission for any of the activities?

☐ Have I decided what types of student groupings I will construct? What will I ask each group to do? How will the groups share information with each other?

☐ Do I understand how to guide the students from activity to reflection?

☐ Have I allowed enough time for the lesson—and extra time in case my estimate is wrong?

☐ Am I certain enough of the lesson's goals so that I can tell when to modify my plan or let the students follow their own ideas?

students' interests, ideas, and questions? Remember the story in Chapter 9 about the students in Ms. Drescher's class who were exploring liquids. In that case, a fairly simple lesson turned into several days' work on density, a concept that was not in Ms. Drescher's original plan.

Some teachers are so pleased with their lesson plans that the plan becomes an instruction manual for the lesson rather than a useful guide. This prevents the teacher from letting go of the plan even when the students' own ideas have taken the lesson to another level of awareness. As you have gathered from the stories in this book, you will often find that your students have questions of *emerging relevance,* to use the term introduced in Chapter 9. You will even have students who ask you directly if they can try a certain experiment that was not in your original plan. More often than not, if appropriate materials are available, it is a good idea to encourage the exploration of these problems of emerging relevance.

Questions of emerging relevance

But how exactly will you know when to let go of your plan? Be assured that careful planning will give you the confidence to let go. This is because you will be able to evaluate the science ideas connected to each topic and consider the possibilities for students to expand their thinking with new activities. And the more experience you gain, the more comfortable you will be in thinking on your feet.

Here are some situations in which you can reasonably modify or abandon your original plan:

Times to modify your lesson plan

■ The science idea that you were hoping would emerge does not.

■ Students reveal alternative conceptions that are resistant to change. (Think about the bottle-and-balloon story in Chapter 3, when the students needed to explore several different setups before they would abandon the idea that the air came from outside the bottle.)

■ The students want to explore a related investigation that you believe is a good idea. (Think about the liquids exploration in Chapter 9, when the students developed an interest in density.)

■ One or more students need to explore a phenomenon on their own terms. (Remember Jamie in Chapter 4, the student who had a different question about icicles.)

It is impossible for you to predict where your students will take an activity that you have planned, but it is important to listen actively to their queries and encourage creative explorations. During the pulling-it-together time, the students can share the particular meanings that their explorations have had for them. The entire class will be enriched by the detours and alternative paths that your students will take.

THE ROLE OF QUESTIONING

Importance of open-ended questions

Often you facilitate the science experience through the questions you ask your students. **Open-ended questions**—those that lead to multiple answers—are especially important because they help students think critically about the investigation. Frequently, too, the teacher's questions lead students toward their own new investigations. As you saw earlier in this chapter, the nature of the questions you plan to ask can form part of your written lesson plan. Now let's take a closer look at the types of questions and the way they can help you guide students.

Types of Questions

Your questioning can be categorized in many different ways. For our purposes, we can divide questions into three main types: those that address process skills, those that help the students focus, and those that challenge the students.

Questions That Guide Students to Use Process Skills

Often you can use open-ended questions to encourage students to apply various process skills. The following list offers samples of questions that promote the use of certain skills. Of course there are many other questions you can use, and many useful ways of phrasing each question.

Sample process-skills questions

Question	Process Skill
What do you notice?	Observation
Why do you think this is happening?	Inference
In what ways are these things the same?	Comparing
In what ways are these things different?	Contrasting
How would you group these objects?	Classifying
What do you think will happen?	Predicting
How would you test this idea?	Planning an investigation
How would you change this experiment if you repeated it?	Planning an investigation

Questions to Focus Students

Sometimes questioning techniques point students toward certain topics, much as the focusing questions at the beginning of the chapters do in this book. In other words, **focusing questions** invite students to come up with their own ideas about a specific topic or investigation. Teachers also use focusing questions to probe their students' understanding of a particular concept related to the investigation.

Sample focusing questions

Here are some examples of focusing questions that relate to the snails investigation described in the sample lesson plan:

- How would you describe the effect of the light on the movement of your snails?
- How would you test how far your snail moves in 20 seconds?
- What evidence do you have for your statement that snails move toward water?
- What type of investigation could you design to explore your snail's reaction to heat?

Three Keys to Good Questioning

1. Ask a question only if you are truly interested in knowing what the *students* are thinking.

2. Design questions to help students construct their own answers.

3. *Never* answer your own questions! It is better to leave a question unanswered for a while than to answer it yourself. The questions you ask the students are always more important than the answers you give them.

Questions to Challenge Students   Often the simplest question can challenge students to think deeply about an experiment or an exploration. My favorite example is, "So, what do you think is going on here?" I use this question to invite students to articulate the meaning they are making of an event.

Questions relating to daily life

We also challenge students when we ask them how the phenomenon they are exploring may relate to something they have seen in their daily lives. For example, with the snails lesson, you might ask your students, "What animals have you seen move like a snail? Are there ever times when people move like snails?"

Another way to challenge students is to ask them simply, "What did you find out from this activity? Can you write it in your science journal?" Or you can ask your students what they liked about an exploration, what they didn't like, and, most important, "How would you change it?" Questions of this type encourage students to think critically. In addition, requiring students to commit their thoughts to writing helps develop their skill in communicating about science and scientific processes. This process parallels the way scientists function as they carry out their research. They make tentative conclusions, try to connect their ideas to other ideas, and record their findings in their notebooks or journals.

Questions for critical thinking and expression

Finally, challenging questions can give you windows into your students' thinking and let you know if they have developed alternative conceptions. Remember that unless you invite students to tell you what they are thinking, a true exchange of ideas cannot take place.

A Word About Wait Time

Whether you are asking students questions in their small groups or addressing the entire class, it is very important to allow students ample time to respond. **Wait time** refers to the time that elapses between the moment when you ask a question and the moment when you select a student to respond, offer a clue, rephrase the question, or otherwise move ahead.

Research findings on wait time

A great deal of research documents the value of providing significant wait time. Studies conducted across grades kindergarten through 12 have compared longer average wait times (generally 3 to 5 seconds) with shorter ones (often 1 second or less). These studies indicate that the longer wait times raise the quality of the student-teacher interactions and the level of the discourse (Chuska, 1995; Rowe, 1974, 1987; Tobin, 1986).

Value of longer wait times

Wait time provides an opportunity for the students and the teacher to think about the exchange and process their ideas. In whole-class exchanges, researchers believe, longer wait time increases the number of students who participate, so that more than just a few students respond. Similarly, in small-group exchanges, longer wait time helps engage all the members of the group.

The literature about wait time has been widely accepted in science education, and yet many teachers continue to rush to call on students. Often this occurs because we are not used to silence in the classroom. Sometimes teachers feel so uncomfortable when students do not respond after 5 seconds that they answer their own questions!

SCIENCE LEARNING GROUPS: CREATING AN ENVIRONMENT FOR COOPERATIVE LEARNING

WORKING COLLABORATIVELY . . . FOSTERS THE PRACTICE OF MANY OF THE SKILLS, ATTITUDES, AND VALUES THAT CHARACTERIZE SCIENCE.

—National Science Education Standards

You may have wondered why so many of the science stories in this book feature classrooms in which students are learning in small groups. The teachers in our science stories created environments that nurtured collaboration among students. The basic reason is that scientific inquiry—the processes of investigation, reflection, and further investigation, whether undertaken by third graders or by adult scientists—benefits greatly from the collaboration of several people.

In this section, we address four important questions surrounding the small-group model of teacher and student interaction:

Questions about
small-group learning

1. How do small-group investigations in science relate to what educators call cooperative learning?

2. In what ways is small-group learning consistent with constructivist views of learning?

3. In what ways is small-group learning consistent with the way science and scientists operate?

4. How can I set up small-group learning in my own classroom?

Cooperative Learning Groups

Research on
cooperative learning

A great deal of important research has been done on **cooperative learning,** which can be defined to mean students' working together in groups to accomplish shared learning goals. Studies have shown a variety of both social and educational benefits from cooperative learning. For example, cooperative learning helps students retain more conceptual knowledge. It also fosters a classroom climate in which students interact with each other in ways that promote each student's learning (Johnson & Johnson, 1994).

What is a cooperative
learning group?

It is important to recognize, though, that cooperative learning, as the researchers in this field define it, is not just any small-group instruction. A **cooperative learning group** is an arrangement in which a group of students, usually of mixed ability, gender, and ethnicity, work toward the common goal of promoting each other's and the group's success. In other words, in a cooperative learning group, each student is responsible for his or her own learning *and* the group's learning.

Formal vs. informal
groups

There are many ways to structure such learning groups. One basic distinction is between formal and informal groups. In *formal* cooperative learning groups, the group stays together until the task is done. In *informal* groups, the commitment is for a shorter term, as when each student checks with a neighbor to see if the neighbor understood (Johnson & Johnson, 1994, p. 100). The science stories in this book have involved formal learning groups, which stayed together until the problem was solved, the model was built, or the task was accomplished.

THE AGE OF
COOPERATION IS
APPROACHING. . . .
TEACHERS AND
ADMINISTRATORS ARE
DISCOVERING AN
UNTAPPED RESOURCE
FOR ACCELERATING
STUDENTS'
ACHIEVEMENT: THE
STUDENTS THEMSELVES.

—*Robert Slavin (1990)*

Constructivism and Small-Group Learning

Cooperative learning groups, such as the ones you have seen throughout this book, encourage the very constructivist learning processes that are at the heart of gaining conceptual knowledge in science. In such groups, students have the opportunity to engage in discussion with others about their ideas, discover differences between their own explanations and others',

In cooperative learning groups, students exchange ideas with other students. Here one student is explaining an assignment to another.

Spencer Grant/Stock, Boston

and defend their positions or alter their thinking as the group strives for consensus. Why is this interaction so important for learning? There are many reasons, but here we concentrate on three: reflection, the social context of learning, and the implications of cultural diversity and gender.

Cooperative Learning and Reflection   Throughout this book, as we have explored the way students learn science best, we have seen that science teaching and science learning rely on the active participation of students. But we have also seen that students must be mentally active as well as physically engaged with activities. In this chapter, by saying that an activity is not a lesson, we have stressed the need for reflective mental processes to accompany the concrete activities.

Small groups foster reflection.

Small-group, cooperative learning fosters these all-important processes of reflection. When students exchange and clarify ideas with their peers, plan an investigation, or refine their observations and conclusions, their minds are constantly engaged. Each student is prodded to more reflection by the inferences and opinions of others. The collaborative problem solver goes further in constructing new ideas and becomes better prepared for higher levels of thinking.

Learning in a Social Context   In Chapter 1, I noted briefly how our understandings of the world are constructed in a social context. Even if you disagree fervently with your neighbor about, say, the value of exploring the moon, you both are working with many similar concepts, including understandings of what and where the moon is and what kinds of exploration are feasible. You are both also using similar language.

In the same way, students' construction of new science explanations and theories is influenced by the social communication and language norms of our society. Peer communication in a small group helps the group members find a common language with which to express their meaning, and this process promotes their learning. For example, in a lesson on density, the textbook might say, "Density is the mass divided by the volume." This type of decontextualized language can be baffling to many learners. But in a small-group conversation, the students may talk about how much 100 milliliters of corn syrup weighs and whether they think the corn syrup is denser than corn oil. One student may then weigh 100 milliliters of corn oil and find out that it weighs less than the corn syrup. As this more contextualized meaning for density emerges from the common social discourse of the small learning group, each student can develop a more personalized, deeper understanding of the science idea.

Cultural Diversity and Gender in Cooperative Learning   As we saw in Chapter 2, images of scientists have traditionally excluded women and minorities. We know, too, that students who see themselves as marginal in a large group do not readily participate. Thus it is not surprising that girls and students of minority ethnic and cultural groups all too often distance themselves during science lessons, contributing little and learning little.

Cooperative, small-group learning activities can help solve the problem. For culturally diverse learners and for girls in science, these activities attend to their academic needs far better than individual learning activities (Barba, 1998). The climate of support created by cooperative learning groups encourages the participation of students who are less likely to volunteer and interact in a whole-class situation.

We should not forget, too, that getting culturally diverse students involved brings benefits to everyone else in the group. Remember Chapter 8's story of Ms. Byrne's third-grade class that was investigating fruits and

---

Sidebar notes:

Shared concepts and language

NOT ONLY DO STUDENTS ACQUIRE KNOWLEDGE FROM EACH OTHER, THEY ALSO LEARN FROM THE PROCESS OF TRYING TO PUT THEIR IDEAS INTO WORDS IN ORDER TO ALLOW SOMEONE ELSE TO UNDERSTAND THEM.

—*Mark Grabe and Cindy Grabe (1998)*

Contextualized meanings

Encouraging diverse students to participate

Benefits for all

vegetables. The students, who were from several cultures, shared information about foods that were familiar and unfamiliar to them. This type of interaction expands everyone's knowledge and often leads students to develop further questions to investigate.

Scientific Investigation and Group Learning

Collaboration by professional scientists

In the adult world, scientific investigation is a social process. Scientific researchers collaborate with one another all the time. They work on teams; they share references to important articles; they access others' research through electronic communication and the Internet.

Eliminating stereotypes, fostering teamwork

Classroom science learning groups are a real-life model of the collaboration necessary for true scientific inquiry to occur. They help eliminate the false stereotype of the lone scientist in a remote laboratory. They promote the kind of interchange and teamwork that are essential for scientific problem solving. Moreover, by fostering interactions among students who normally would not be relating to one another, they widen the range of students' observations and increase both the breadth and the depth of classroom investigations. The more diverse the perspectives are on nature, the more possibilities there are for thorough explorations of natural phenomena. In addition, scientists agree that the more diverse the nature of observations and inferences is, the greater is the possibility that an accurate idea will emerge in the conversation.

Structuring Cooperative Learning Groups in Your Classroom

There are many different ways of structuring small learning groups. No matter how you do it, however, some questions will always arise: Which students should be in which groups? What individual roles need to be assigned within a group? And how should I, as a teacher, mediate learning for the various groups in my class?

Balancing group size and diversity

Assigning Students to Groups Most small learning groups are created with two to four students. Sometimes groups of five are effective, but usually the smaller, the better. Keeping the group size small may limit the amount of diversity you can achieve within each group. Nevertheless, try to design the groups with as much heterogeneity as possible. Students of mixed ability, gender, and ethnicity should have the opportunity to work together toward common goals, learning about one another in the process.

Problems with student
choice of groups

After all this book has said about involving students in the scientific process, you may think it wise to let students select their own groups. I do not recommend this practice. Student-selected groups are often homogeneous, with males working with males, white students working with other white students, and so on (Johnson & Johnson, p. 104). Students do not then interact with a wide variety of peers, one of the goals of cooperative learning.

Seeking student input

You may want to gather information from the students before you design the groups. For example, you can ask each student to write down three names of people he or she would like to work with. With this information, you can build groups that include at least one person of choice for each member. Groups should work together for as long as it takes for them to be successful, but at some point you will want to reshuffle the groups so that each student has the opportunity to work with as many other students as possible.

Importance of roles

Assigning Group Roles   In formal learning groups, each member usually is assigned a particular job or role. This is important not just to promote group efficiency but to make sure that every student participates fully. When everyone has a specific task essential to the problem solving at hand, each student will be a valued member of the group.

Sharing responsibility
between teacher and
students

In the model of science learning groups that I prefer, the teacher designs the groups, but the members of the group then select a leader, who is usually the spokesperson, responsible for reporting results to the entire class. This leader changes from week to week or project to project. The teacher also specifies other roles to be filled, and the group selects individuals to fill them. For example, one person could be in charge of getting materials. Another member might be the recorder, writing down the group's plans as the ideas are formulated and then making the notes available to the entire group for their science journals. Another student could be responsible for group clean-up, and still another might be the checker—the person who checks for understanding, asking all members what they think.

The more complex the investigation and the larger the group, the more roles may be involved. Whatever the assigned tasks for one investigation, the roles should change for the next project.

Your role as mediator

Mediating Group Learning   As the teacher, you manage and mediate the student groups. You give directions for establishing the jobs and assigning the tasks, and then you facilitate the change of jobs when a new investigation begins. You can promote effective group work by monitoring how the group is functioning and making suggestions. Reminders to individual

Helpful Hints for Planning a Science Lesson for Student Groups

- Keep in mind that an activity is not a lesson. The lesson encompasses the thinking, reflection, and personal connections that occur because the students were engaged in the activity.

- Be yourself, and be comfortable with what you and your class are doing.

- Get yourself out of the way. That is, instead of delivering lockstep instruction, do the following:

  Provide experiences.

  Observe your students as they explore.

  Gain insight into how they work together.

  Listen to their ideas.

  Offer suggestions, ask questions, prompt, and coach.

- Prepare and plan as much as you can. Then be prepared to alter your original plan if the class's own ideas generate a new direction, one that you may not have thought of before.

- Remember that your lesson plan should guide you, not limit you.

students like, "Did you check with your group?" or "What does your group think?" can help promote the value of the group (Johnson & Johnson, 1994, p. 102).

Different groups will make different decisions about how to proceed with scientific investigations, and it is important to honor and value these **Honoring differences** differences. Sometimes it will even be necessary to allow an individual to go off on an investigation by himself or herself, as Mr. Wilson did with Jamie in Chapter 4. At the same time, however, you need to coach and prompt the groups as they struggle toward problem solving. When appropriate, offer ideas, provide focus, or give specific explanations of the problem at hand.

When the investigation is complete, invite the members of each group to discuss how well they collaborated with each other. Their comments will help you evaluate the groups' effectiveness and plan for the next group science activity.

As you create this model in your own classroom, you will find that small-group cooperative learning is truly essential to doing science with children. You will see for yourself that the groups encourage the process of problem solving in ways that individual instruction cannot.

QUESTIONS FOR YOUR OWN REFLECTION

Recording your reflections

When the science experience you have planned for your students is over, you may want to document what went on. If you use a constructivist approach, there is always a lot to write about because you really do not know beforehand where the lesson will go or what the children will say or do. It is the children's thinking that propels you forward. Besides actively listening to the students during the lesson, it is useful to take some time afterward to record their ideas and your own reactions to the way the lesson developed.

Questions to reflect on after the lesson

Here are some sample questions to ask yourself as you reflect on the lesson and write about it in your science journal:

- What did the students find out in this experience? Were there any surprises?
- How did the students in each group work together? Were there any problems?
- Was the activity open-ended enough, or did each group do more or less the same thing?
- How did the students extend the investigation?
- How did the students connect this experience to their daily lives?
- Overall, what do I think the children got out of this experience?
- What do I remember most about this science activity?
- Would I do it again?
- How would I plan differently the next time?

Save your plans and notes!

Sometimes it is useful to record comments and reflections directly on the lesson plan itself. Keeping your plans together in a notebook is a good idea too. Your comments and notes have important implications for how you will address the topic the next time.

Using a computer as an aid

If you are comfortable using a computer, it can help you not only to create your lesson plans, but also to save them and add comments later. In the next chapter, we explore other ways in which computers and technological aids can facilitate your planning, your research, and your students' activities.

Key Terms

lesson *(p. 255)*
lesson plan *(p. 255)*
open-ended questions *(p. 262)*
focusing questions *(p. 263)*
wait time *(p. 265)*
cooperative learning *(p. 266)*
cooperative learning group *(p. 266)*

Resources for Further Exploration

Electronic Resources

Pitsco's Launch to Lesson Plans. **http://www.pitsco.inter.net/pitsco/lesson.html**. A web site full of links to lesson plans, many of them relevant to elementary school science.

Welcome to IMSEnet. **http://www.ncsu.edu/imse**. From the Network of Instructional Materials for Science Educators, this web site offers invaluable hypertext links to lesson plans and activities. See Chapter 13 for additional web resources for lesson plans.

Print Resources

Cohen, E. G. (1994). *Designing Groupwork Strategies for the Heterogeneous Classroom.* New York: Teachers College Press.

Eltgeest, J. (1985). The right question at the right time. In W. Harlen (Ed.), *Primary Science: Taking the Plunge.* Portsmouth, NH: Heinemann.

Freedman, R. L. H. (1994) *Open-Ended Questioning: A Handbook for Educators.* Menlo Park, CA: Addison-Wesley.

Johnson, D. W., &. Johnson, R. T (1994). *Learning Together and Alone: Cooperative, Competitive, and Individualistic Learning.* 4th ed. Boston: Allyn and Bacon.

Slavin, R. E. (1990). *Cooperative Learning: Theory, Research and Practice.* Englewood Cliffs, NJ: Prentice Hall.

Wilen, W. W. (1991). *Questioning Skills for Teachers.* 3d ed. Washington, DC: National Education Association.

Elementary School Science and Technology: A Seamless Connection

FOCUSING QUESTIONS

- What comes to mind when you think of technology?
- What types of experiences have you had with computers and computer applications?
- How do you use the Internet? For personal use? Professional use?
- What are your favorite web sites?

I have always loved to play the board game Scrabble. A few months ago, a friend gave me a computer software program that plays Scrabble with me. The board appears on the screen, I automatically receive seven letters at random, and then it is my turn to play. After I put a word on the screen's playing board, the computer displays my total points and tells me the number of tiles I need to replace. The computer then makes up a word from its own selection of letters, trying to earn enough points to beat me.

It is not difficult for a skilled Scrabble player to beat the computer occasionally. Unlike the real game, the computer program gives hints, if requested, about which words to play. This simulation has greatly improved my performance in a real Scrabble game. I love practicing with the computer because it is private, I can get help if I need it, and the computer doesn't sigh loudly when I take too long for my turn.

When I realize how useful my computer program has been in improving my Scrabble game, it makes me think in general about educational uses for good computer software applications. There are many computer resources—with more being developed each day—that teachers can introduce to students in the course of their elementary science experiences. Programs designed mainly for practice, like my computerized Scrabble, are among the least ambitious of computer software designs.

You may remember that in Chapter 2 of this book I talked about "locating your scientific self." You will not be surprised to learn that the same type of advice holds true for technology as well. Getting comfortable, in any way that is personally relevant to you, is a first step toward integrating the use of technology in your classroom.

In this chapter, we explore many different possibilities for using technology for the teaching and learning of elementary school science. We chiefly look at the use of computer applications for instructional purposes. We will examine the ways that computer technology can facilitate both science as a process and science as a set of ideas. First, though, consider what the term *technology* means in science education.

THE MEANINGS AND USES OF TECHNOLOGY

Many people assert that human beings were technologists long before they were scientists. The first person who fashioned a spear from the slender branch of a tree was modifying the world to meet a set of needs for himself or herself. That, according to some definitions, is the essence of technology. It is broadly thought that the goal of science is to *understand* the natural world, while the goal of technology is to *modify* the world to meet human needs (National Research Council, 1996, p. 24). Or we could phrase the distinction in this way: science usually explores that which has always been there, while technology creates that which has never been before.

Comparing technology to science

Over the course of time, though, science and technology have become deeply intertwined. As humans explored the natural world, they became increasingly dependent on the technological tools that they developed to assist them. Eventually the tools of scientific inquiry—the microscope and telescope, for example—provided a great deal of data about the natural world, and that in turn helped technologists design even more sophisticated tools.

A reciprocal relationship

Today, science and technology are closely related. In fact, they have a reciprocal connection. The need to know more about the natural world prompts new technological developments, and new technology then drives scientific research, promoting new understandings about the natural world and new questions, which then demand new technology to provide answers. A good example of this reciprocal connection is the 1969 lunar landing. Scientific knowledge of the moon and its orbital relationship to earth, coupled with a deep desire to explore the moon, spurred development of the Apollo rocket and the Lunar Exploration Module. That technology, which allowed human beings to step on another celestial surface for the first time, brought back new data about the moon that stimulated further scientific research and further advances in technology.

Dual nature of technology

With this connection in mind, you can see that technology, like science, has a dual nature. Technology is both a process and a product. The process of technological design has given us products, ranging from pencils to laptop computers, that have become tools for instruction and tools for learning. These **instructional technology tools** and their applications in the elementary science experiences are the central topic of this chapter.

Integrating Technology with Instruction and Learning

Technology in daily life vs. techology in schools

Central to this chapter is the belief that technology should be fully integrated into the experience of learning and teaching science, so that the connection between the two becomes seamless. This is true in our private lives, where technology has become so pervasive that we scarcely notice it.

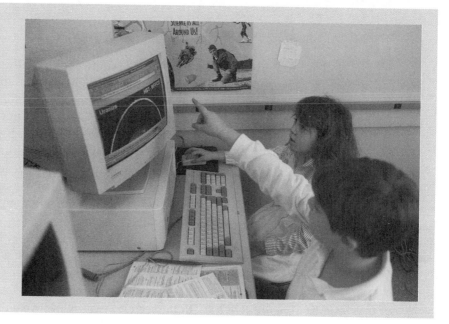

Computers are excellent resources for scientific data. Here, students are exploring the Internet for data about Uranus.

C.W. McKeen/The Image Works

When you place a telephone call, send a fax, or type an email message, you spend little or no time pondering the technology itself; you just use it. Unfortunately, technology rarely plays the same natural role in classrooms (Grabe & Grabe, 1998). Much remains to be done to integrate technology into our schools.

Why make the connection "seamless"?

But why, you may wonder, should the use of technology in classrooms be "seamless" and "natural"? Why doesn't technology, quite exciting in its own right, provide its own science lesson to students? The reason relates to the goals of the science lesson that we discussed in Chapter 12. Remember the question, "What am I hoping the children will get out of this science experience?" The response to this question implies the fuller understanding of one or more science ideas and the use of several process skills. So if you want your students to learn about the solar system, for example, and to practice using their skills of observation and inference, you do not want them to become preoccupied with accessing a web site with wonderful pictures of the planets. Rather, you hope they will use all available resources, including the computer, to support their investigation.

When we aim for a seamless connection between school science and technology, we are asking ourselves this fundamental question:

In what ways can technology help my students and me satisfy our goals for meaningful science instruction?

The Need to Know

At the heart of the process of scientific inquiry is the *need to know*. Whether the students are figuring out how to light a light bulb, or wondering if their garden snails are ever going to move, or noticing the tiny green leaf that finally emerges from a seed, they are invested in the outcomes. In order to continue the investigation or confirm some findings, they need to know something, and this need creates its own momentum.

Technology as a resource for building meaning

It is in this context that technology can help the learner. It can make the search for information easier, supply another example, connect a student to others who are doing the same investigation, provide data, or help to begin a new inquiry. Computer technology and its related resources—the Internet, software packages, interfacing laboratory instruments, and multimedia presentation packages—help learners construct their own meaning. Many computer technologies allow independent, learner-directed uses that fit comfortably into the learner-directed nature of inquiry-based science activities. In the sections that follow, we will examine a number of ways in which technology can help satisfy the need to know.

MAKING OBSERVATIONS AND GATHERING DATA

This section explores ideas for using computer and video technology to enhance the scientific processes of observing and collecting data. In doing so, technology can also support the processes of making inferences, creating models, and planning investigations. Every day new technologies improve the computer's ability to act as a "lab assistant" for scientific investigation. The applications described below are examples of the ever-increasing possibilities.

Obtaining Information from the Web

Real-time data

In one of the science stations in Chapter 5, the students were asked to access a weather site on the Internet in order to obtain real-time weather data in three cities and make comparisons. Clearly, this exploration would be impossible without computer technology. There are many other cases, too, in which elementary school science data can be obtained from World Wide Web sites.

Fostering a community of learners

Imagine, for example, that a third-grade class is tracking the path of monarch butterflies as they migrate in early fall. This annual migration is recorded on databases available on the web, and the data are public information, as are many types of earth and space science data. In the case of monarch butterflies, the third graders could find a web site where they could contribute their own data and read other people's reactions to viewing the migration. This type of experience—a community of scientists and casual observers alike sharing data via a central site—is unique to the Internet.

Another example might involve a fifth-grade unit on volcanoes and earthquakes. Real-time updates on volcanic activity are available on specific web sites. In one class I observed, the students were exploring volcanic activity in their own state and three neighboring states over the past one hundred years. They learned that even areas not known for volcanic eruptions sometimes have considerable tectonic activity deep below the surface.

Scientific agencies on the web

Often students' spontaneous curiosities propel them to seek data from the web. (Later in this chapter, we will examine criteria for evaluating the reliability of web sites.) Most scientific agencies and public information systems, including the U.S. Wildlife Service, the National Audubon Society, the U.S. Geological Survey, and many other reputable institutions, publicize on their web sites the data they collect as they track birds and wildlife, earthquakes, and volcanoes. These data become excellent resources for students engaged in collecting information for their own investigations.

Data for Making Models As we observed in Chapter 10, building models often requires gathering data about the object or process we want to represent. The fourth-grade students in that chapter, who were making a model of the solar system, used the Internet as one of their resources.

One of my favorite models, you may remember, was my own model of the tooth. Recently I went on the Internet and did a web search for "teeth." What an extraordinary number of images and explanations were available for my tooth model! If I were building my model today, I would surely want to use some of that Internet information.

In another fourth-grade class I observed, the students were asked to create a project that would demonstrate their understanding of photosynthesis. One student made a "talking leaf" as her project. Karen brought in a cardboard model of a green maple tree leaf with a tape recorder attached to the lower surface. When she played the tape recorder, it gave an explanation of the process of photosynthesis using the first-person voice, as if the leaf itself were speaking. It turned out that she had done her research on the Internet, and when I asked her why, she said she had convenient access to it and there were lots of sites to choose from.

Online information for creating models

Information from Working Scientists A number of web sites offer contact with scientists who will answer questions about specific topics in a variety of disciplines. Often, too, these sites contain archives of past questions and answers. Several such web sites are listed in Table 13.1. Besides supplying information, these sites give students the sense of interacting with an actual scientist, even if only in cyberspace.

Sometimes students and teachers can use the web or email to establish more personal connections with scientists. When one fifth-grade class in the East was studying invertebrates, the teacher dug up 20 worms from her backyard garden. With a collection of large jars and potting soil, each group of students constructed a home, or wormery, for these invertebrates. As the students examined their earthworms, each group came up with a different set of questions to explore.

One group was concerned about whether their observations would be true for all types of earthworms. How did their worms compare to other earthworms in different parts of the country? Their teacher encouraged them to explore the World Wide Web to find out. Doing so, the students lo-

Direct connection with a scientist

TABLE 13.1 Web Sites for Asking Questions of a Scientist

Name of Site	URL Address	Description
Ask a Scientist	http://scorescience.humboldt.k12.ca.us/fast/ask.htm	Allows users to ask questions of scientists working within a variety of disciplines (e.g., geologists, physicists)
Ask a Scientist!	http://www.hmsc.orst.edu/education/whales/aska.htm	The Hatfield Marine Science Center of Oregon State University will answer questions about whales. The site also offers information about whales in aquariums and species that are endangered.
Ask the Space Scientist	http://image.gsfc.nasa.gov/poetry//ask/askmag.html	An astronomer answers approximately 30–50 questions per day. Archives of past questions are sorted by topic.
MAD Scientist Network	http://www.madsci.org	More than 250 scientists from around the world answer questions. The site also allows students to perform online activities and research.
WhaleNet	http://whale.wheelock.edu	Research scientists with the Whale Conservation Institute of Wheelock College, Boston, will answer questions about whales. Hundreds of questions from other users have been archived on this site.
Windows to the Universe	http://www.windows.umich.edu	Allows students and teachers to communicate with astronomers.

cated a professor in Utah who had been studying earthworms of a different variety. From the professor's web site, the students found out how their worms compared to the Utah worms, and then they emailed the professor their own data about worms.

Evaluating Information from the Web The ability to evaluate the information that you and your students discover on the Internet is critical. You can find material on the World Wide Web that you cannot find in any other medium—partly because any computer-savvy individual can post content on the web. The upside of this situation is the remarkable access we now have to original and creative data. The downside is that there is no gate-keeping for the material that makes its way onto this infinite communication network.

The following points can help as a general guide in evaluating web sites:

Guidelines for evaluating web sites

1. If you find a web page that has information you want to use, learn who the author or organization is.

2. Enter the author or organization on a search engine, and see what you find. If nothing shows up, watch out! That often means the source is not reliable.

3. Check out the hypertext links on a web page. These links should point to other, external sites for your exploration. If all the links point back to internal pages, you may have unreliable data.

4. Communicate with the author of the web site by email if you cannot verify the source's information. Reliable web page authors will respond and give you additional information. If the authors do not respond, that could be a sign that the information is suspect.

Instruments with Computer Interfaces

Interfacing instruments are laboratory instruments that come with a means to connect them to the personal computer. The most common are probes with related software. The software package is usually on a disk that can be inserted in the computer disk drive, and the probe itself usually has a connecting plug or computer card that fits into a slot on your computer. In computer terminology, the probe "interfaces" with the computer.

You may be wondering, "Why bother with such additional, fancy equipment?" The reason is that a computer probe can measure properties that are difficult to measure with conventional devices in the classroom. In one third-grade classroom, the children were exploring whether heat is produced during seed germination (see also Reynolds & Barba, 1996, p. 58). They were germinating seeds in plastic bags with wet paper towels and had inserted a temperature probe into one of the bags. This probe took the temperature of the air inside the bag every 2 hours over the course of 5 days. Collecting the data every 2 hours in a conventional way would

Extending the ability to collect and record data

have been impossible, since no one was around at night. Moreover, the probe stored the data on the software disk, making them easily accessible and manipulable. Using this equipment, the children collected data that indicated a steady temperature rise as the seeds germinated.

Other students in the same class became further involved with the temperature probes and seed germination. Before they left one Friday afternoon, they set up four temperature probes—three in three different seed bags and one in the open air of the classroom—so they could compare the air temperature outside the bags with the temperature inside. On Monday morning, the students were waiting eagerly at the classroom door to see the results.

Video Cameras: Extending Our Senses

Science teachers are finding that a video camera and a tripod are useful technologies for improving and extending the process of observation. One science room teacher set up a video camera and a slow-running videotape the night the class hamster was having babies. Another teacher used a video camera on a tripod to tape the popping of a water balloon. Watching the videotape in slow motion, the students saw things they could not have seen with the unaided eye. Their videotape helped them answer the question, "What really happens when a balloon bursts?" (With a water balloon, from which the water is expelled as the balloon pops, the bursting process is slow enough for the details to be revealed in a video.)

Learning from teacher-made videos

Another excellent use of the video camera is the *video-microscope*, which reveals a microscopic world on a classroom television set. Using this technology, one fourth-grade class observed the single-celled organisms in a small sample of water from a nearby pond. The water was set up on a slide of a microscope connected to a video camera, which displayed its images on a television screen. Because the technology allowed all the students to view the slide at the same time, the teacher was able to facilitate a conversation about the microorganisms with the entire class.

Microscope images on a video screen

EXTENDING STUDENT COLLABORATION THROUGH EMAIL

In everyday life, electronic mail (email) has become a common means of communication. It also offers a great potential for extending student collaboration. As you have seen in earlier chapters, communication of ideas is an integral part of doing science. It is a way to deepen and refine one's understanding of a science concept. Email enlarges and facilitates this communication and crosses geographic boundaries in the process.

Email buddies

In one application of email collaboration, students in an elementary classroom have email "buddies" in another class across the country. Typically, the buddies' exchanges are related to a common unit or theme. For example, monarch butterflies have different migration patterns on the East Coast and the West Coast. Children tracking these patterns on opposite sides of the country can communicate with each other about what they have learned. In this way, the students get a new perspective on the science topic from their buddies' points of view. In addition, the contact with people from other regions and other cultures often leads to greater understanding of diverse lifestyles.

Other possible uses of email

Students and teachers continue to invent new ways of using email. Here are some of the possibilities (Grabe & Grabe, 1998, pp. 195–196):

- *Global classrooms,* in which several classes study a common topic and exchange accounts of what has been learned.

- *Electronic appearances,* in which a scientist can respond through email to students' questions, as if the scientist were making a guest appearance in the classroom.

- *Electronic mentoring,* in which students in younger grades may be paired with mentors in older grades for email tutoring and exchange of ideas.

USING COMMERCIAL SOFTWARE FOR ELEMENTARY SCIENCE INSTRUCTION

There are many different types of instructional software with elementary science applications. As with the exploration of the World Wide Web, the use of computer software should arise from a need to know generated by the students' questions and investigations.

Interactive, problem-solving software

The software packages that best lend themselves to inquiry-based activities are those that pose a problem and engage students in interactive manipulation of a simulated natural event. You may wonder why you should use a computer simulation instead of a regular, hands-on experiment in the classroom. The reason is similar to that for using models, which we explored in Chapter 10. That is, computer-simulated investigations often focus on the exploration of phenomena that cannot be studied directly in the classroom.

Software dissection

As an example, the use of live animals for dissection has become a highly controversial issue. For this reason, among others, it is rarely attempted at the elementary science level. However, software simulations of dissections, just like simulations of the Scrabble game, are very useful. They are designed to be interactive—the student acts, and the simulated environment reacts. As the student conducts the electronic procedure, more and more of the inner organs are revealed.

Other simulations available on computer diskettes or CD-ROMs allow students to learn in a controlled way about the motions of the planets in the solar system and how electrons flow though an electrical circuit. Many other elementary science subjects also lend themselves to the use of interactive software. Besides permitting students to observe phenomena that are not readily available in schools, simulations often allow manipulations of the environment in ways that would not be possible in the real world.

Evaluating Software

For any simulation to be useful in creating personal meaning, it has to be interactive with the students in ways that foster their own problem solving. How will you know if a software package can help students build their own meanings? Consider these criteria for evaluating computer software:

Criteria for evaluating
software

- Does the software engage the students in a problem-solving activity?
- Does it guide students to think about one or more new science ideas?
- Are there opportunities for the student to personalize the experience presented by the software?
- Does the software portray gender or ethnic stereotypes?
- Is the language engaging and at the students' learning level?
- Are the instructions clear?
- Are the graphics and design clear and uncluttered?
- Is there a high degree of student involvement as a result of using this software application?
- Does the program encourage further discovery learning when it is over?

Where to find software
reviews

Grabe and Grabe (1998) recommend looking up software reviews for commercial products in magazines for educators. For elementary science software, the *Journal of Computers in Math and Science Teaching* and *Computers in the Schools* carry software reviews. The National Science Teachers

TABLE 13.2   Web Sites Offering Reviews of Educational Software

Name of Site	URL Address
California Instructional Technology Clearinghouse	http://tic.stan-co.k12.ca.us
Children's Software Revue	http://zippy.tradenet. childrenssoftwarenet
Software Publishers Association Education Market Section	http://www.spa.org/project/resource.htm
Technology & Learning	http://www.techlearning.com
Thunderbeam	http://www.thunderbeam.com

Association, based in Arlington, Virginia, publishes its own software reviews, as does its elementary science magazine, *Science and Children.* There are also web sites devoted to software reviews, such as the ones named in Table 13.2.

RESOURCES FOR TEACHERS

Not only does technology offer rich resources for students' investigations, it also offers many aids for the teacher. In this section we will look at two of them specifically: elementary science lesson plans that you can find on the World Wide Web, and the uses of technology in communicating with other teachers.

Lesson Plans on the Web

A wealth of lesson plans

The World Wide Web offers a wealth of elementary science lesson plans. These plans range from the simple to the extremely ambitious, from small-group and single-class projects to multiclass collaborative affairs. The more you explore them, the more ideas you can garner for your own use.

You can find lesson plans on some of the web sites listed in the Expanding Meanings sections throughout Part Two of this book. In addition, Table 13.3 lists a number of sites that I have found useful, for both general information and lesson plans. I am not suggesting that you should adopt a lesson plan exactly the way you find it at a web site. Rather, you can examine various plans, then pick and choose among their ideas or modify them for your own classroom.

Evaluating Lesson Plans   For any lesson plan or suggested student activity you find on the web, you may want to apply the general guidelines for web site evaluation already listed in this chapter. If the source looks unreliable, be skeptical of its ideas for students.

Beyond that, here are some specific guidelines for evaluating lesson plans, based on the lesson plan format presented in Chapter 12:

Guidelines for
evaluating online
lesson plans

- Does the lesson plan articulate clear learning goals for the students?

- Does it explain the science ideas behind the lesson?

- Is the plan written for a specific age or grade level?

- Does the lesson plan indicate materials and where to get them? (Note this especially for live specimens.)

- Does the lesson plan give lockstep instructions for the student to follow, or does it provide for flexibility in a constructivist learning environment?

- Does the plan make connections to the students' lived experiences?

- Does the plan indicate related sources, literature, or activity books?

Electronic Communication with Other Teachers

The Internet offers many possibilities for communicating with other future teachers. As you explore the world of science and children, getting in touch with others at similar stages of their careers can help you discover yourself as a teacher and work out answers to some of your puzzles.

In one recent example, preservice teachers from New York wondered what they would have in common with future teachers from another part of the country who were also struggling with how to do science with children. Through a joint project that their professors set up, using the Internet to establish email partners, the New York students were able to communicate with future teachers from Boise, Idaho, and Central Queensland, Australia. After receiving each other's email addresses, the far-flung partners began to engage in conversations about the following questions:

Teachers as email
partners

TABLE 13.3   Some Resources on the World Wide Web for Elementary Science Educators

Name of Site	URL Address	Description
Earth & Sky Online	http://www.earthsky.com	An interactive site with excellent links and access to scientific communication, based on a daily science radio series
Global SchoolNet Foundation	http://www.gsn.org	Emphasis on global collaborative learning projects
The GLOBE Program	http://www.globe.gov	A collaborative, Internet-based environmental education project
Hands-on Science Centers Worldwide	http://www.cs.cmu.edu/afs/cs/usr/mwm/www/sci.html	Links to public museums that emphasize interactive science education
The Learning Studio @ the Exploratorium	http://www.exploratorium.edu/learning_studio/index.html	Online and other activities from San Francisco's Exploratorium museum
The Learning Web	http://www.usgs.gov/education	Activities and teaching resources in earth science, from the U.S. Geological Survey
Newton's Apple	http://ericir.syr.edu/Projects/Newton	Experiments and lesson plans connected with the popular television program
Online Educational Resources	http://quest.arc.nasa.gov/OER	A NASA site that offers links to many resources related to space exploration
PBS Online Science	http://www.pbs.org/science	Educational activities and resources associated with PBS television programs
Frank Potter's Science Gems	http://www-sci.lib.uci.edu/SEP/SEP.html	Links to science activities arranged by topic and grade level
Scholastic	http://www.scholastic.com	Activities arranged by subject and grade level
The Smithsonian Institution Lesson Plans	http://educate.si.edu/lessons	Lesson plans keyed to the Smithsonian's exhibitions

Finding an Email Partner

As a future teacher, you can look for an email partner through the listserve of the Association for the Education of Teachers in Science. Go to the association's web page (**http://science.coe.uwf.edu/aets/aets.html**) and read the instructions about becoming part of the electronic list. Many science education professors have found each other through this listserve, setting up email partnerships with each other and for their future teachers.

You can also enter chatroom conversations on the Internet that explore science education. One example is the Learning Network chats about educational projects, which you can find at **http://www.sln.org/schools/index.html**. You may chat with the educators there or join their ongoing collaborative projects. You can also chat with visitors to the site about their work and your own. Similarly, you can find a link to chat sessions at the web site called *Teachers Helping Teachers* (**http://www.pacificnet.net/~mandel**).

1. How do children learn science best?
2. What is the prevailing image of the scientist?
3. What are some local issues that can inform science teaching?

Many preservice teachers were surprised that, despite different geographic locations and local interests, their science teaching philosophies proved remarkably similar. The following excerpts are from the email dialogues of the student pairs:

Email dialogues of preservice teachers

Rachel and Cindy (Rachel's notes)

In communicating with my partner, Cindy, I found out that we share many ideas about teaching science. Cindy stressed that science should be "hands-on" and fun. I challenged her about the "fun" part and said I didn't think that teachers should get so carried away with "fun" that they lose their main purpose along the way. The students should be active in situations that require some type of thinking; they must be given an activity that makes them think and problem-solve in an open way! We both agreed on the value of tying the material in the classroom to one's life. This makes everything you do more useful and valuable to the children.

Jennifer and Iris

JEN: *What kind of learning environment do you think children learn science best in? I feel that hands-on along with minds-on is the best. It just takes a lot more work on our parts as teachers.*

IRIS: *As far as children learning science, I agree with you. Hands-on and minds-on. If students don't experience things for themselves, they are not going to remember or even understand the idea. Secondly, they will not know how to apply it to their lives and past experiences.*

Such email experiences encourage preservice teachers to discuss their roles as science teachers. They often find that these electronic chats with peers are invaluable.

YOUR ROLE IN PROMOTING COMPUTER TECHNOLOGY

As you begin your teaching career, you should find out the following information about your school:

Questions to ask about your school . . .

- How many computers are available to the students on a daily basis?
- Are the computers linked to a communication network?
- Do the students have easy access to the World Wide Web?

This information will help guide you in planning computer interactions for your students. Of course, the way you encourage computer use will also be determined by your answers to questions about your students' investigations:

. . . and about your students

- What are the students' questions?
- What type of data are they looking for?
- How can the computer help them find the answers?
- What relevant software packages are available for class use?

Facilitating the integration of technology in science lessons is a part of doing good science. Your role here is essentially the same as your role in facilitating meaningful science experiences. When the technology deepens and enriches the science experience—and it often does—encourage and

IF COMPUTERS ARE
DESTINED TO PLAY AN
INCREASINGLY
IMPORTANT ROLE IN
EDUCATION OVER THE
NEXT 20 YEARS, IT IS
NATURAL TO ASK WHAT
ROLES WILL BE PLAYED
BY HUMAN BEINGS.

*—President's Committee of
Advisors on Science and
Technology (1997)*

coach its use. Help your students use technological resources as an integral and natural extension of their senses as they search for answers to their scientific questions.

This chapter has only scratched the surface of the topic of instructional technology and its uses. Chapter 15 will discuss uses of technology for assessing students and for your own record keeping. But most of what you learn about technology will come—and should come—from your own experiences with it as you begin to teach. Remember, the guiding principle for using computer technology in the elementary science experience must be, "Does it fit? Is there a need it can meet?" Chances are it *will* fit and it *will* meet many of the students' needs to know.

Key Terms

instructional technology tools *(p. 277)*
interfacing instruments *(p. 282)*

Resources for Further Exploration

Electronic Resources

Classroom Connect. **http://www.classroom.net**. A web site that offers many resources for teachers, including links to other sites devoted to science teaching and technology integration.

Learn the Net. **http://www.learnthenet.com/english/index.html**. A useful site for learning how the Internet and the World Wide Web work.

Teacher/Pathfinder. **http://teacherpathfinder.syr.edu**. The Schoolhouse section offers resources in science and technology, among other subjects. The site also has information on assessment and professional development, plus an area called The Teachers Village that features ideas about lesson plans, field trips, classroom management, and more.

Print Resources

Dublin, P., Pressman, H., & Barnett, E. (1994). *Integrating Computers in Your Classroom: Elementary Science.* New York: HarperCollins. This useful handbook includes helpful hints about using software as well as lesson plans in which computer applications are the central activity.

Gilster, P. (1997). *Digital Literacy*. New York: Wiley. This book's title refers to the ability to understand and evaluate the multitude of information, in a multitude of formats, that we get through computers. A fine discussion of the need for digital literacy in contemporary society.

Grabe, M., & Grabe, C. (1998). *Integrating Technology for Meaningful Learning*. Boston: Houghton Mifflin. A detailed and helpful guide to using technology in an integrated manner.

Learning and Leading with Technology. Eugene, OR: International Society for Technology in Education. Published eight times a year, this magazine features ideas for integrating technology with instruction.

Reynolds, K., & Barba, R. (1996). *Technology for the Teaching and Learning of Science*. Boston: Allyn and Bacon. This book speaks directly about the unique attributes of science teaching that are congruent with technology applications.

Elementary Science Curriculum: From National and State Standards to Your Classroom

14

FOCUSING QUESTIONS

- What is the difference between a book of science standards and a curriculum?
- Who decides what science topics to teach at each grade level?
- How do I use the National Science Education Standards in my own science teaching?

A new teacher who is planning to implement a meaningful science program in the elementary classroom may find the number of curriculum guides, standards documents, and science activity books staggering. You may wonder, "What do I *do* with all these materials? How can I use them to create a science program in my own class?" This chapter explores the role of curriculum guides and standards and helps you determine how to make good use of them.

When I was a brand-new teacher, another teacher in my school gave me the science textbook and said, "Here, this is the curriculum you need to cover." I wondered, then and for some time after, "How can a textbook be a curriculum?"

Eventually I learned that schools have been very busy "covering" curriculum for many years. Often this coverage is so exacting that another meaning of the term *cover* comes into play—that is, the science ideas and the skills that students might be expected to attain from elementary science programs remain hidden! "Covering" curriculum too often means memorizing facts from the textbook, which is considered the final authority.

In *Science Stories*, you find a contrasting picture. As teachers, our role is not just to "cover" a curriculum, but rather to help children *uncover* the ideas and meanings that emerge through their experiences. Our approach in *Science Stories* is dedicated to the idea that knowledge is constructed by learners; it cannot simply be transmitted from a text to the learner. This constructivist perspective does not diminish the authority an expert text writer might possess. It does, however, stress the idea that the ultimate authority for any person's learning is situated within that person. In this approach, the science textbook has a less important role; it becomes a resource, not a curriculum. Some schools, in fact, do not even use textbooks, because teachers and students rely on trade books and the Internet for scientific information.

"Well," you may be wondering, "if I don't use a textbook to guide me, what *do* I use?" The following sections describe what is meant by a curriculum and the types of materials that you can use as resources.

WHAT IS A CURRICULUM? ....

The word *curriculum* derives from the Latin term meaning "running course." The term has evolved over time to represent a program of studies for a school, a grade level, or a particular class. Even this understanding has multiple interpretations. For example, some educators think of curriculum as the content of a course of study. Others view curriculum as a plan of studies that includes both content topics and types of teaching activities. Still others view curriculum as including everything—all the formal and informal activities that take place in a school or a classroom.

Multiple
interpretations of
curriculum

In elementary science, we usually think of the **curriculum** as a plan of studies that includes the ways in which the science content is organized and presented in each grade level. It is not a list of topics, but an organizational scheme that describes the activities that facilitate the presentation of the content. Often schools use a list of topics as a *summary* of the science curriculum. (See Table 14.4 later in this chapter for one such summary.)

Curriculum as an
organizational scheme

Along with this *formal* science curriculum, as we mentioned in Chapter 6, good teachers make use of an *informal* curriculum based on the students' daily lives. The informal curriculum includes everyday, often unexpected, topics that arise as a result of living in a given geographic environment or exposure to a newsworthy science topic. Often formal and informal science curriculum topics merge. For example, if a fourth-grade class in a seaside community is studying the oceans as a formal unit, the teacher and

The informal
curriculum

students can find many opportunities to incorporate aspects of their daily lives into the formal study.

WHO CREATES THE CURRICULUM? NATIONAL INFLUENCE AND LOCAL CONTROL

In American public education, matters of curriculum have traditionally been the prerogative of local school districts. The last two decades, however, have brought a great deal of national concern about the competencies and academic performance of students. As we mentioned in Chapter 1, the publication in the 1980s of reports like *A Nation at Risk* and *Educating Americans for the 21st Century* alerted the public to the perceived need to address higher standards in American public education. The subsequent "standards movement" has resulted in national standards for several areas of education, including science.

Growing national concern

To understand the forces involved in establishing curriculum, let's look first at two sets of national standards that have affected many states and local communities.

The *Benchmarks* and the *National Standards*

Science education has been particularly influenced by two major national reform efforts, represented by two documents that are useful tools for fashioning science curriculum: the *Benchmarks for Science Literacy* (1993), a publication of the American Association for the Advancement of Science, and the *National Science Education Standards* (1996), a publication of the National Research Council.

Both documents represent the work of teacher educators, scientists, and classroom science teachers from all over the world. The intent is to provide educators with a way of thinking about the importance of scientific literacy in today's world. **Scientific literacy** refers to an individual's ability to use scientific information to make choices and to "engage intelligently in public discourse and debate about important issues that involve science and technology" (National Research Council, 1996, p. 1). Both documents argue that such literacy should be the major goal of science education.

Scientific literacy defined

Benchmarks is a product of Project 2061, a comprehensive initiative by the American Association for the Advancement of Science to improve science education. (The project is named for the year Halley's Comet will

Goals of Project 2061

next reappear.) The project's first publication, *Science for All Americans* (Rutherford & Ahlgren, 1990), delineated science literacy goals for all students and stimulated efforts to develop science programs that emphasized fewer science concepts but greater depth of exploration. *Science Stories* has modeled that approach by implicitly emphasizing depth rather than breadth. *Benchmarks*, Project 2061's second publication, includes statements of what students should know and be able to do in science, mathematics, and technology by the end of grades 2, 5, 8, and 12.

Shared themes

The National Science Education Standards (NSES) are aligned with the standards set out in *Benchmarks* in their conceptualization of science content at each grade level. The NSES include principles guiding the content, methods of teaching, and assessment of school science, as well as standards for science education programs and for science education in large school systems. An overarching theme in both the NSES and the *Benchmarks* is that a sound science program needs to engage students in inquiry—that is, real problem-solving and design experiences that help them shape meaning from experience. Table 14.1 summarizes some of the general emphases in the NSES, emphases that are also reflected in the *Benchmarks*.

TABLE 14.1   Curriculum Emphases of the National Science Education Standards

Less Emphasis	More Emphasis
Knowing scientific facts and information	Understanding scientific concepts and developing abilities of inquiry
Studying subject matter disciplines (physical, life, earth sciences) for their own sake	Learning subject matter disciplines in the context of inquiry, technology, science in personal and social perspectives, and history and nature of science
Separating science knowledge and science process	Integrating all aspects of science content
Covering many science topics	Studying a few fundamental science concepts
Implementing inquiry as a set of processes	Implementing inquiry as instructional strategies, abilities, and ideas to be learned

From National Research Council, *National Science Education Standards.* Washington, DC: National Academy Press, 1996, p. 113.

State and Local Interpretations

Individual states have also become involved with science curriculum decisions. Some states have provided their own documents, sometimes referred to as **frameworks,** that offer guiding principles for elementary and secondary science curricula. Recent state frameworks have usually been aligned with the content standards of the NSES. These state frameworks tend to dictate content standards—the science topics that should be addressed at each grade level.

Although the national standards and state frameworks have a great deal to say about what science curriculum should look like, local schools and school districts still have the final word in defining their own curricula. The standards movement should be seen as a guide to influence the development of curriculum. None of the standards or frameworks documents is a curriculum in itself.

Keep in mind, too, that topic selection is only part of the story of curriculum construction. To develop a curriculum, educators also need to explore related activities, such as connections with literature, mathematics, technology, and social studies. Furthermore, they need to evaluate ways in which the informal curriculum can link to the content of the formal curriculum.

A few years ago, I saw a fifth-grade geology curriculum designed by a school in a shore community of Long Island, New York. It had an extensive section on sand, including required reading about the sands of the Sahara Desert. This reading segment was part of the elementary science textbook that had been purchased by this school. Certainly, the Sahara, being the world's largest desert, was of interest, but it was unlikely that students could explore it directly. Unfortunately, the curriculum offered the students no direct access to sand, even though there was a beach full of it—directly accessible and relevant—down the block from the school.

This story illustrates the importance of using many criteria to design a curriculum. Keep it in mind as we examine the science content standards of the NSES. Also remember the key question, "Who are my students?" Whatever the formal curriculum, answering this question will help you make connections to the lived experiences of your students.

Frameworks

Framework ≠ curriculum

What is missing here?

THE EIGHT CONTENT STANDARDS OF THE NSES

Eight content
categories of the NSES

The *National Science Education Standards* present eight categories of content standards, which together encompass science as both a set of ideas and a set of processes:

1. Science as inquiry

2. Unifying concepts and processes in science

3. Science and technology

4. Science in personal and social perspectives

5. History and nature of science

6. Physical science

7. Life science

8. Earth and space science

Let's look at what each of these categories includes.

Making local connections enhances the experience of science in your classroom. Teachers can use standard curriculum and modify it to match the personal lives of the students. Here children are planting in a nearby vacant lot.

David Young-Wolff/Photo Edit

Science as Inquiry

The science-as-inquiry standard is the basis for the other seven. Basically, inquiry in elementary school science refers to the following:

- Engaging students in the processes of science.
- Engaging students in analyzing the results of investigations and re-thinking their plans.
- Helping students to understand the difference between evidence and explanation.
- Guiding students to develop an understanding of the recursive nature of scientific exploration.

The stories in Part Two of this book showed you what science-as-inquiry looks like in practice.

Unifying Concepts and Processes

The unifying-concepts standard expresses the need for students to understand that every topic in science is probably part of a larger system or process that will help us understand how the natural world works. This standard reminds us to ask ourselves, "What is the bigger picture to which this topic is attached?"

For an illustration, think back to the snails lesson from Chapters 8 and 12. This topic can be seen as part of a larger, unifying concept in science education that the standards label "form and function." By studying their land snails, the students will begin to understand how the snails' shape and physical attributes (form) relate to the snails' behavior (function). The NSES identify four other major unifying concepts, which are listed in Table 14.2 along with relevant science stories from Part Two. Often, of course, science activities and explorations are linked to more than one unifying concept. Table 14.2 identifies only a few of the links you might find in Part Two's science stories.

Science and Technology

The science-and-technology content standard calls for students to use the skills involved in designing a solution to an identified problem and then implementing and evaluating the solution. **Design,** in this sense, is the technological counterpart to the science-as-inquiry process. The difference

TABLE 14.2   Important National Science Education Standards Unifying Concepts and Processes Linked to Science Stories in Part Two

Unifying Concept	Brief Explanation of Concept	Relevant Science Stories in Part Two
Systems, order, and organization	Since the natural world is very large and complicated, scientists typically organize their investigations in terms of systems, that is, groups of related objects or organisms.	Exploring Solids, Liquids, and Gases (Chapter 7) Mysterious Matter (Chapter 7) States of Matter and Scrambled Eggs (Chapter 7) What Does It Mean to Be Alive? (Chapter 8)
Evidence, models, and explanation	Scientists collect observations and data to use as evidence in formulating their explanations. Often they use models as aids, especially when they cannot observe an object or phenomenon directly.	Investigating a Natural Disaster (Chapter 6) The Cans of Soda (Chapter 9) The Floating Egg (Chapter 9) Floating and Sinking Fruits (Chapter 9) An Edible Solar System (Chapter 10) A Model Orbit (Chapter 10) Shapes of the Moon (Chapter 10) Batteries, Bulbs, and Wires Revisited (Chapter 11)
Constancy, change, and measurement	Understanding a system typically involves (1) identifying and measuring its changes over time and (2) discovering the aspects that remain constant over time.	A Snow Story (Chapter 6) A Tree Grows in the Bronx (Chapter 6) States of Matter and Scrambled Eggs (Chapter 7) Looking at Liquids (Chapter 9)
Evolution and equilibrium	For both living and nonliving systems, understanding the present often involves learning how it evolved from the past.	Two Seashores (Chapter 6) From Seed to Plant (Chapter 8) What's Inside a Seed? (Chapter 8) Planting in a Vacant Lot (Chapter 8)
Form and function	The form of an object or living being frequently relates to its function.	The Osage (Chapter 6) When Is a Vegetable a Fruit? (Chapter 8) A Book of Snails (Chapter 8)

Adapted with permission from *National Science Education Standards*. Copyright © 1996 by the National Academy of Sciences. Courtesy of the National Academy Press, Washington, DC.

is that the design process results not only in deeper understanding, but also in a concrete product.

Again, we can find examples in the science stories of Part Two. In Chapter 9, students designed and constructed science display toys using liquids of different densities. Their problem was to use the liquids they had been exploring to create a new kind of toy. Similarly, in Chapter 11, some students designed and constructed a shoebox house, while others built a blinking lighthouse or a model flashlight. These students were tackling the problem of applying what they had learned about electrical circuits to new purposes. In both stories, students designed a technological solution to a particular problem, and they had the opportunity to establish criteria for their own optimum solutions. They also faced the constraints on what they could do, a useful component in the learning process.

> **CHILDREN'S ABILITIES IN TECHNOLOGICAL PROBLEM SOLVING CAN BE DEVELOPED BY FIRSTHAND EXPERIENCE IN TACKLING TASKS WITH A TECHNOLOGICAL PURPOSE.**
>
> —*National Science Education Standards*

Science in Personal and Social Perspectives

Learning applied to personal, social concerns

The standard concerning science in personal and social perspectives calls on science education to help students connect their learning to personal and social issues. The standard also states that students should develop decision-making skills that will aid them with the everyday problems we all face as citizens of a modern society. I think of this standard as forging the types of connections to students' lives that I illustrated in Chapter 6. In the elementary schools, this standard relates to local science and technology issues such as recycling, waste removal, environmental impacts of human actions, and changes in populations and personal health.

The standard has a strong and important social studies component. We can see that component in the earthquake story in Chapter 6, when students became aware of the causes and social implications of a natural disaster. Overall, this standard calls for students to explore the impacts of science and technology on their lives as citizens and to be deeply aware of their natural environments.

History and Nature of Science

The standard for history and nature of science addresses science as a human endeavor. It reminds us to teach students that science is a body of knowledge created by groups of women and men using certain skills and employing certain processes to study the natural world.

As you have seen throughout this book, understanding the nature of science is as important as understanding a particular science concept. In fact, the two are inseparable. Hence, in the NSES standard, understanding "the nature of science" means understanding science as inquiry—how science actually works in practice—as well as grasping the use of process skills in science. This approach is based on research indicating that unless we teach the nature of scientific activity directly, students do not learn about it (Lederman, 1992).

The nature of science is closely linked with its history. Using this standard as a guide, you might want to explore, with your students, the ways in which scientific ideas have changed with time.

Suppose that your students are building a solar system model like the one in Chapter 10. In this model, they will place the sun as the point around which the nine planets and their satellites move. In this context, you might want to extend the lesson to incorporate historical beliefs about the solar system. For centuries, people believed that the earth was the point around which the sun and all the planets traveled, until Copernicus, a Polish scientist who lived and worked in the 1500s, suggested that the earth is actually hurtling through space around the sun. Learning about the pre-Copernican view of the solar system can help students realize that science is an ongoing human activity that continually produces new un-

. . . .

TABLE 14.3   **National Science Education Standards for Physical Science, Life Science, and Earth and Space Science**

	Physical Science	Life Science	Earth and Space Science
Grades K–4	Properties of objects and materials Position and motion of objects Light, heat, electricity, and magnetism	Characteristics of organisms Life cycles of organisms Organisms and environments	Properties of earth materials Objects in the sky Changes in earth and sky
Grades 5–8	Properties and changes of properties in matter Motions and forces Transfer of energy	Structure and function in living systems Reproduction and heredity Regulation and behavior Populations and ecosystems Diversity and adaptations of organisms	Structure of the earth system Earth's history Earth in the solar system

Table 6.8 and Table 6.9 taken from National Research Council, *National Science Education Standards.* Washington, DC: National Academy Press, 1996, pp. 109–110.

derstandings. It is useful, too, to remind students that alternative conceptions can hang around for a very long time.

Physical Science, Life Science, and Earth and Space Science

Standards for traditional subject areas

These three content standards, listed in Table 14.3, address widely accepted divisions of the domain of science. They focus on the science facts, concepts, principles, theories, and models that students should know in each broad area. Besides matching the content to appropriate developmental stages and learning abilities of students, these standards attempt to supply guidelines that are genuinely usable in developing science curricula. Later in this chapter, you will see how these standards can be applied to a sample curriculum.

MST (MATHEMATICS, SCIENCE, AND TECHNOLOGY) CURRICULUM AND STATE FRAMEWORKS

Blending math, science, and technology

In addition to connecting science firmly with technology, the NSES discuss coordinating the science program with the mathematics program (National Research Council, 1996, pp. 214, 218). Recently, in fact, a great deal of attention has been given to the links among math, science, and technology. One movement in curriculum transformation, which we can call the **MST initiative,** involves truly integrating the disciplines of mathematics, science, and technology in curriculum planning. MST stresses the fact that mathematics, science, and technology are interdependent human enterprises. It promotes science literacy by bringing technology education—the entire concept of engineering design—into the direct support of both science and mathematics activities. It also stresses the use of inquiry and design in problem-solving activities.

This initiative is making an important impact on the development of elementary school curricula. In New York State, for example, elementary teachers are urged to explore their elementary science curriculum and "MST" it. That means reviewing the problem-solving investigations with respect to the following criteria:

MST criteria

- What is the nature of the problem-solving activities in science?

- Where does mathematics present itself in these activities?

- Are children expected to design a process or a product as part of their investigations?

- What is the nature of students' access to computer information systems during their investigations?

- How are science ideas linked to mathematics and technology ideas?

Examples of MST

In Part Two of *Science Stories*, you saw several examples of elementary MST investigations. In Chapter 9, when Ms. Drescher's students used their understanding of the differing densities of three liquids to design a liquid display toy, all three components of problem solving—mathematics, science, and technology—came into play. In Chapter 10, when students designed a model solar system, they used their analysis of planets' relative sizes and distances to construct a model that accurately represented the planets in relation to the sun. The students also accessed the Internet to gather data about the nine planets. As a final example, some of Ms. Travis's students in Chapter 11 designed a shoebox house and planned the wiring for ceiling lights in the rooms. They used the science ideas behind series and parallel circuits to light their model house, and they used mathematics to construct model furniture in an appropriate size.

DESIGNING ELEMENTARY SCIENCE CURRICULA ....

After reading about standards and initiatives, you may still be puzzled about how schools and teachers get from such broad principles to specific curriculum plans. To see how this might be done, let's explore the science curriculum topics for one school district in the Northeast. Table 14.4 presents an outline of this district's curriculum.

From standard . . .

To begin, consider how the school district applies the NSES standard for life science curriculum. This standard for grades K–4 (Table 14.3) lists "characteristics of organisms" and "life cycles of organisms." The NSES also include the following explanation of these topics:

During the elementary grades, children build understanding of biological concepts through direct experience with living things, their life cycles, and their habitats. These experiences emerge from the sense of wonder and natural interests of children who ask questions such as: "How do plants get food? How many different animals are there? Why do some animals eat other animals? What is the largest plant? Where did the dinosaurs go?" An understanding of the characteristics of organisms, life cycles of organisms, and of the complex interactions among all components of the natural environment begins with questions such as these and an understanding of how individual organisms maintain and continue

TABLE 14.4   Topics for the K–5 Science Curriculum in a Sample School District

	Life Science	Physical Science	Earth Science
Kindergarten	"All About Me" (the five senses) Farm animals Seeds and plants (planting gardens)	Classifying: buttons, apples	Weather and seasonal science: daily readings
Grade 1	Ourselves Living/nonliving: ■ Pets, animals, and habitats: a wormery ■ Egg to chick (life cycles)	Properties of matter: ■ Solids, liquids, gases ■ Sink/float	Objects in the sky
Grade 2	"Terrific Trees" Caterpillar to butterfly (life cycles) Seed to plant (life cycles): germination bags	Light and color Electricity and magnets "Looking at Liquids"	Dinosaurs
Grade 3	Plants: classifying fruits and vegetables Animals: ■ Classifying vertebrates and invertebrates ■ Mealworms ■ Owl pellets The human body	Water cycle and weather Sound energy Heating and cooling Simple machines	Rocks and soil Earthquakes Planets and constellations Milk waste and conservation Recycling paper
Grade 4	Food chains and webs: local habitats and ecosystems	Chemical and physical change Heat energy	Environmental science: ■ Oil spills ■ Air and water pollution Earth motions
Grade 5	Plants and animals: ■ Tropisms ■ Five kingdoms ■ Vertebrates and invertebrates	Solutions and suspensions Light energy: reflection and refraction Electricity and magnetism: ■ Parallel and series circuits ■ Electromagnets	Recycling and conservation: local issues Moon phases

life. . . . Children's ideas about the characteristics of organisms develop from basic concepts of living and non-living. (pp. 127–128)

. . . to curriculum

The school district addresses this standard with various life science topics ranging from farm animals in kindergarten to food chains and webs in fourth grade (see Table 14.4). Each topic relates to and reinforces a general understanding of life cycles and characteristics of living things.

Notice that some of the topics repeat themselves, more or less, as the students move through grades K–5. For example, the students explore seeds and plants in both kindergarten and second grade. They study a particular animal's life cycle in first grade and again in second grade. During the unit on owl pellets in grade 3, they examine what an owl eats; then in grade 4 they address the more general subject of food chains. This **spiraling of curriculum,** as it has been called (Bruner, 1960), is one way to develop depth of understanding of a topic. Spiraling does not mean, of course, that students engage in the same activity over and over. Rather, as they mature, they build on earlier science experiences and develop a greater depth of understanding. You can see a similar spiraling in the physical science column of Table 14.4: for example, electricity and magnets in grade 2, and then electricity and magnetism with rather complex subtopics in grade 5.

Spiraling

Developing Curriculum Units

In Chapter 12 I drew the usual distinction between a lesson and a unit. Typically a **unit of study** includes several lessons designed around a central theme or topic. In Table 14.4 each of the entries would probably represent one unit of study. An individual lesson on snails, for example, could fit within the third-grade unit on classifying vertebrates and invertebrates.

Elements of a unit plan

A unit plan usually consists of the following:

- The science ideas behind the unit
- The activities included in the unit
- The lesson plans for the unit

Sometimes unit plans also include a description of the assessment strategy for each unit. We will address several types of assessments in the next chapter.

Steps from content standard to unit plan

Aligning a Unit of Study with the Standard   Developing a unit of study from the guiding principles in a content standard involves the following steps:

1. Identifying the topic and its relationship to the standard.

2. Identifying the science ideas behind this topic.

3. Exploring activity books and other resources relating to the topic.

4. Designing lessons that allow children to explore their own ideas and permit the teacher to extend the activities as needed.

5. Indicating connections with mathematics, technology, literature, and social studies.

Selecting Activities   If you are selecting activities for a unit, you should follow the same criteria described throughout this book for individual lessons. Ask yourself questions like these:

Questions for selecting unit activities

- Does the activity lend itself to individual thinking and problem solving?

- Does the activity match the science idea expressed by the unit?

- Are the materials accessible?

- Is the activity student centered or teacher centered?

You are looking for science activities that encourage individual explorations of phenomena. As we observed in Chapter 11, commercial materials and kits can be useful as long as you frame the activity in a way that allows students to make decisions about the plan of their investigations.

Activity guides

The Resources for Further Exploration at the end of this chapter offer some starting points for selecting activities. The activity guides listed there have an array of topics that explore all the areas in life, physical, and earth and space science. And as you gain experience, you will come across your own activity books as well as Internet resources.

What Is Missing?   When you rely solely on a national standard to guide your school's or your class's elementary science curriculum, you miss a great deal of the local influence that ought to contribute to the students' science experiences. Who your students are, how they experience nature, where they live, what region of the country your school is located in—all of these considerations should help you frame your units of study.

Incorporating local and personal links

Remember, too, to explore the personal lives that your students live. Are they hungry? Poor? Overindulged? Disabled? Middle class? How will you use science curriculum to help students to make meaning in the context of their own lives? That is the major role of your science curriculum: to engage students in their own learning, relate science experiences to the students' lives, and help students explore and draw conclusions. Certainly, national documents should help you understand the range of topics and ideas that are appropriate. But what is best for *your* students remains a local decision.

Curriculum—any curriculum—is a lifeless document. You, the teacher, give it life when you use it to guide your students' experiences.

A CHECKLIST FOR THE SCIENCE CURRICULUM

In evaluating a science curriculum, here is a checklist that you can use:

Questions for
curriculum evaluation

- What is the role of scientific inquiry in the curriculum?
- Are there earth science, life science, and physical science topics at each grade level?
- Are technology and mathematics incorporated into the science activities?
- Where are connections to other subjects, such as social studies and literature?
- Do the topics spiral from lower to higher grades?
- Are there topics that make local geographic connections and have personal relevance for the students?
- Are the topics explored in depth?
- Is there room for informal science experiences?

If you are involved in creating the curriculum, these questions should be answered along the way. If you as a teacher are working with a curriculum designed by others, these questions can help you identify its strong points and weak points and decide when and how to supplement the formal curriculum with informal experiences.

Key Terms

curriculum *(p. 294)*
scientific literacy *(p. 295)*
framework *(p. 297)*
design *(p. 299)*
MST initiative *(p. 303)*
spiraling of curriculum *(p. 306)*
unit of study *(p. 306)*

Resources for Further Exploration

Electronic Resources

Educational Standards and Curriculum Frameworks for Science. **http://putwest.boces.org/StSu/Science.html**. Maintained by Charles Hill and the Putnam Valley Schools in New York, this site offers an annotated list of Internet sites for K–12 educational standards and curriculum frameworks, including a breakdown of frameworks by state.

Marco Polo. **http://www.mci.com/marcopolo**. Developed by MCI, the American Association for the Advancement of Science, and the National Geographic Society, among other groups, this web site offers activities and lesson ideas keyed to *National Science Education Standards* and *Benchmarks for Science Literacy.*

National Science Education Standards. **http://www.nap.edu/readingroom/books/nses**. This web site includes an online version of *Standards.*

Project 2061: Science Literacy for a Changing Future. **http://project2061.aaas.org**. Describes Project 2061 and includes an online version of *Benchmarks for Science Literacy.*

Print Resources

American Association for the Advancement of Science. (1993). *Benchmarks for Science Literacy.* Washington, DC: The Association.

Kessler, J. (1996). *The Best of WonderScience: Elementary Science Activities.* Albany, NY: Delmar Publishers.

National Research Council. (1996). *The National Science Education Standards.* Washington, DC: National Academy Press.

Curriculum Activity Books and Guides for Elementary Science

Great Expectations in Mathematics and Science: The GEMS Series. Available from the Lawrence Hall of Science, University of California, Berkeley, CA 94720-5220; (510) 642-7771.

Improving Urban Elementary Science: Insights. Available from Education Development Center, 55 Chapel Street, Newton, MA 02160; (800) 225-4276.

Science and Technology for Children: The STC Series. Developed by the National Sciences Resource Center. Available from Carolina Biological Supply Co., 2700 York Road, Burlington, NC 27215; (800) 334-5551.

What's the Big Idea?
Assessing What Students Know and Are Able to Do

FOCUSING QUESTIONS

- How do you know if the students "got" the science idea?

- What do you think it means to "match assessment to instruction"?

- How can a "performance" be an assessment?

- What role can technology play in assessment?

When I was in southern Texas working with a group of third- and fourth-grade teachers, they invited me to spend some time with their students. I was interested in learning what the students thought about science—how they would define it, for instance, and how they felt about learning it in school. One day I brought a tape recorder to class and interviewed the students about what they thought science was. This is how it went:

ME: What do you study in science?

THIRD-GRADE BOY: (really thinking) Scienzz, scienzz—what we're learning about scienzz. . . . There are stop scienzz, one-way scienzz, and yield scienzz.

My northern accent was clearly unintelligible to this student. He thought I had asked about "signs."

This story reminds me of the many occasions when we are trying to find out what students know but the student misunderstands the very way we ask or write the question. We, as teachers, are seeking one kind of meaning, and our students, with the best intentions, offer another kind. How, then, can we design ways to understand what students really know and are able to do in science?

This is the central theme of this chapter. Teachers refer to this quest to determine students' understanding with terms like *assessment* and *testing*. Let's explore what these terms mean in theory and in practice.

ASSESSMENT AND TESTING

Assessment is an activity that teachers and, in fact, everyone else, engage in. We constantly find ways to assess ourselves and others. In education, though, this term is used in specific ways.

Assessment involves collecting information.

Science assessment refers to a process of collecting information that is used to determine the quality and character of an individual or group performance in a science learning experience. As you might imagine, this process of collecting information about what your students know and are able to do includes many different techniques. As two experts in the field put it, "When we assess students, we consider the way they perform a variety of tasks in a variety of settings or contexts, the meaning of their performances in terms of the total functioning of the individual, and the likely explanations for those performances" (Salvia & Ysseldyke, 1998, p. 5).

MANY CURRENT SCIENCE ACHIEVEMENT TESTS MEASURE "INERT" KNOWLEDGE— DISCRETE, ISOLATED BITS OF KNOWLEDGE— RATHER THAN "ACTIVE" KNOWLEDGE— KNOWLEDGE THAT IS RICH AND WELL-STRUCTURED.

—*National Science Education Standards*

Testing is a narrower term. It refers to the use of teacher-made tests as well as state and local tests designed by educational agencies or testing services. Even within these categories, as you are probably aware, tests come in an almost infinite variety. The phrase *paper-and-pencil tests* generally refers to tests made up of short-answer or multiple-choice questions. Essay tests, in contrast, require longer responses involving students' reflective thinking. Although essay tests may be taken with paper and a pencil, the term *paper-and-pencil test* does not usually mean an essay test.

Paper-and-pencil tests

Fragmentary knowledge vs. constructed meaning

Short-answer and multiple-choice tests tend to assess small pieces of knowledge, asking students to recall some term or fact related to the unit of study. Typically, these tests assess knowledge in a fragmented way, rather than in the context of the students' "learning." As we noted in Chapter 3, recall of a term or a piece of knowledge is quite different from constructing meaning. It should not surprise you, then, to learn that paper-and-pencil tests are not the best method of assessing the type of deep, contextual scientific learning that is addressed in this book. To assess that deeper level, educators have been turning to performance assessment.

In **performance assessment,** also known as **authentic assessment,** students demonstrate their understanding by solving a problem in the

Performance assessment: Using the "need to know"

real-life context of their classroom or their world. As you have seen throughout this book, the act of doing science always occurs in context.

Science is not isolated bits of observing, inferring, comparing, or recording; it is instead a contextual whole in which skills are employed because of a need to know. The skills are not important solely for their own sake. In the same way, a performance assessment sets up a need-to-know scenario that encourages the student to employ the skills the assessment is intended to measure.

Assessment that matches instruction

Performance assessments, as you will see in this chapter, can take a variety of forms. Whatever form is used, it should be appropriate to the context in which you are teaching. In other words, it has to *match the instruction*. In fact, a good assessment usually looks like a good instructional task. For example, as we teach science, we engage students in science journal writing. The journals can then also be used for assessment.

Finally, I like to use the term **evaluation** to refer to the process of making value judgments, based on the results of assessments, about a student's or a group's achievement in a science learning area (Doran, Lawrence, & Helgeson, 1993). When teachers evaluate their students' progress, they examine the assessment results and then make a judgment about how well the students understand a concept and use their science skills. Typically, evaluations relate students' progress to their own prior performance and to that of their peers within the class.

Value judgments

Today, many educators define this type of evaluation as part of assessment. In fact, when you hear the term *assessment*, it may even include the decisions that teachers go on to make after they have gathered their information and made their value judgments. For instance, deciding whether a student or a class should move on to the next unit could be considered a final step in assessment. In practice, be aware that assessment is a multi-faceted process, and the word can mean different things to different people.

So much for definitions and qualifications. The question remains, How do we find out what our students know and are able to do in a given area of science? The next stage in answering that question is realizing that assessment is not separate from instruction. Rather, the two join naturally together in the instructional context.

ASSESSMENT AND THE INSTRUCTIONAL CONTEXT

A family relationship

Whenever we are doing science with students, we are engaged in assessment. We wonder, "How are they doing? Do they work well together? What science skills are they using? Are they solving a problem? Do they get the science idea? What kind of model are they constructing?" Assessment and instruction may be thought of as siblings. They "live" in the same classroom, have the same parent (you, the teacher), and interact with the same people in their daily lives. Assessment is an integral part of instruction.

Spotting alternative conceptions

The context of your own classroom will provide many opportunities to assess the students. More often than not, you will use several assessment strategies to learn what the students understand about a science idea as well as what they understand about doing science. As you noticed in Part Two of this book, understanding a science concept often takes time, and there are alternative conceptions along the path to full understanding. You need to know where the children are in their thinking about an idea before you proceed to the next idea. Hence, you are always assessing as a natural part of creating meaningful science experiences.

If you follow the constructivist approach to instruction recommended by this textbook, you will want to explore your students' understanding in terms of their ability to employ knowledge to make sense of a situation or a problem. In other words, you will be looking at their learning in context, rather than just their "possession" of a certain item of knowledge. The following sections discuss several useful techniques: journals, portfolios, conversations, drawings, as well as other types of performances. We will also look at methods for using technology in assessment.

USING SCIENCE JOURNALS FOR ASSESSMENT

Student journals vs. teacher journals

There are many ways to invite students to record a description of science ideas and activities in which they are engaged. I encourage the use of **science journals,** which Chapter 2 described as a personal account of science experiences. Student science journals typically have a more specific structure than the science journal you may be keeping. The structure directs students to the type of information they need to provide. Within this structure, though, the students' personal ideas about their science experiences have plenty of room to emerge.

A Sample Structure for a Science Journal

One way to structure an elementary science journal is to use a list of questions as a guide. Students respond to the questions in writing and by drawing pictures. Here is a sample list of questions:

1. What do I know about _____?

Expressing prior knowledge

For a particular lesson, you would fill in the blank with an appropriate word, such as "snails," "electricity," or "seeds." This question is one for the students to answer *before* they begin a particular investigation or experiment. In this phase of journal writing, they express their prior knowledge; they process their existing ideas and understandings.

For example, in the snails lesson, a student might write the following:

Snails are like slugs with shells. Snails move slow. They live in the dirt.

2. What am I trying to find out?

Defining the problem

This phase of journal writing prompts students to define one or more problems relating to the investigation. The students express the main problem they are trying to solve, formulate any other questions they may have, and perhaps think about extensions to the primary activity. By extensions to the primary activity we mean ideas that come up as they formulate the question. For example, in this section, a student might write:

How do snails move? What do they do when you shine a light on them? What happens when you put water near them? What happens when you put a piece of food near them?

3. What materials do I need?

Listing the materials

Here the students list their materials. In the snails lesson, for example, they would note a flashlight for testing the snails' response to light, lettuce for testing the response to food, and so on.

4. What did I do?

Recording the activities

Once they have answered the first three questions, the students proceed with their investigation. Afterward, in question 4, they describe their own activities. In some ways this section is like a log. It provides students with a record of what they did that led to their understandings. It can be written in a series of steps or in a narrative form.

5. What happened?

A record of findings

In this phase of journal writing, the students document their observations. In the higher elementary grades, this section may be a formal record

The duck eggs have hatched! Joyfully a first grader holds the baby duck. Soon she will write about this life cycle and her journal will be used as a form of assessment.

Mary Quinn

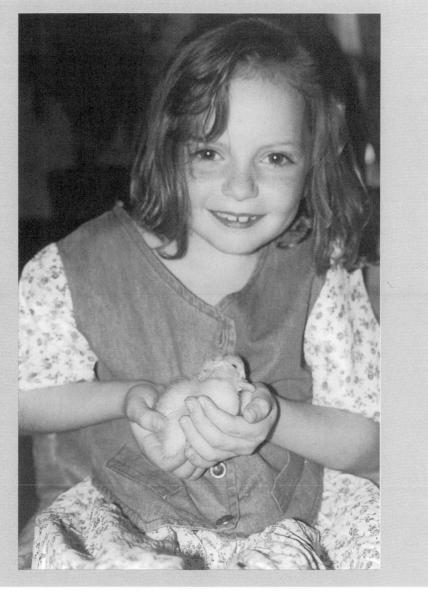

of findings, and the students may arrange their data in chart form or in graphs. In earlier grades, when the teacher has done a demonstration, this "what happened" section relates to what the class and the teacher have noticed together.

Even after a demonstration activity, the science journal should express the *individual* student's experience with the event. Students may draw or write their observations, or both. Figure 15.1 illustrates first graders' journal records of three ducklings' hatching in the classroom and what happened when they were taken outside by the teacher.

Inferences and understandings

6. What did I find out?

In this section of the journal, children express the inferences and understandings they have arrived at as a result of their investigation. For example, in the snails journal, students might write:

The snails do not like the light.
The snails like to eat.
My snail moved very slowly.

For an example combining drawing and writing, see Figure 15.2.

Expressing opinions

7. What do I think about this experiment [or investigation, or activity]?

In this section, students offer their opinions or ideas about the investigation. Sometimes they make connections to the real world—for example,

I like this experiment. It reminds me of how my mom says I move just like a snail.

Planning modifications

8. If I did this over again, what would I do differently?

This section invites students to analyze the investigation and plan modifications. Sometimes they start the investigation all over again. This analysis and documentation of their thinking often leads them to important findings. After the snails lesson, one student wrote:

Next time, I would try this experiment by putting drops of water on the table. My snail didn't do anything when I put the water out in a dish.

Variant journal forms

Overall, a structure of this sort guides students' science writing and their science thinking. Journal entries do not always have to take the same form, however. Sometimes you might choose a more general question to prompt your students' thinking. For instance, during a unit on electricity, you might ask them to write in their journals on the topic, "What I wonder about electricity is. . . ."

A stimulus for further thinking

Besides encouraging students to keep a record of their science experiences, a science journal integrates their personal voice with their observations and inferences. Adult scientists maintain notebooks that become important sources for their research. Architects, artists, inventors, writers, and others keep journals and sketchbooks that record their ideas and observations and inspire further discoveries. In the same way, science

FIGURE 15.1

....

Observations from the science journals of first graders.

(a) A drawing on a teacher-prepared page.
(b) Transcription of a handwritten journal entry.

Announcing...

.... the arrival of our duckling

Date of birth: <u>May 29, 1997</u>
Place of birth: <u>incubator</u>
Room: <u>C7</u>
Duckling's Name: <u>Dudley</u>

Picture at birth

Yesterday mrs. quinn pot the ducklings out side   they were running arwnd like crasy   when mrs. quinn was potting the ducklings back in the jar they were running away from her   they wantied to stay out in the outside world   they like the outside world alot

Reprinted by permission.

FIGURE 15.2
. . . .

A page on "the water
cycle" from a young
student's science
journal.

The sun evaporates the water. The
water droplets form clouds. When
the clouds are hevey, they rain and
the sun evaporates the water agen.

journals do more than record students' emerging science ideas; they
stimulate additional thinking and investigation. And when students look
back over their science experiences as they have described them in their
journals, they feel a considerable sense of pride.

Integrating Technology with Science Journals

If science journals are written by hand, they should always be kept in a bound notebook, not as separate sheets of paper. As an alternative, though, like other creative writing products, science journals can be created with a word processing program.

Advantages of computer journals

Students who keep their science journals on a computer disk can easily add, update, and correct. Moreover, if they have ready access to the Internet or CD-ROMs, they can often insert pictures and other images. For the teacher's assessment of student knowledge and learning, the journals are usually printed out as hard copy and pages are stapled together.

Drawing Pictures and Telling Stories

Students can record their observations in drawings as well as in words. Often I ask students to draw pictures in their journals of what they have observed or experienced during a science activity. Drawing gives them an opportunity to express themselves in a modality other than the verbal-analytic, and this can be an important consideration in our classrooms of increasingly diverse learners.

Importance of nonverbal expression

This form of assessment is not limited to the early grades. In fact, older students are often pleased and challenged when invited to draw what they have experienced. Older students usually place captions on their drawings and explain their understandings in writing as well.

Need for specific criteria

Remember that we are not telling the children *what* to draw. Rather, we are creating an opportunity for them to express their own ideas through the drawing. Nevertheless, we have to know why we are asking them to draw and what we expect their drawings to contain to demonstrate their understanding. In other words, we need to establish criteria for judging the students' performance, just as we would with any other assessment technique.

A clash of expectations

In this context, I'm reminded of a third-grade teacher who took her students on a "tree walk" and then asked them to draw a tree and label its "most important" parts. Her minimum expectation was that the students would properly identify trunk, leaves, bark, and roots. Many of her students, however, did not draw the roots because they were underground. Instead, they drew the trees as they had seen them on the tree walk. This clash of expectations can be avoided when the teacher establishes assessment checklists that match the context of the students' experiences.

With young children, it is useful to have them tell a story about their drawings. As they elaborate and evaluate what they drew on their paper, the teacher gets an excellent idea of what they know about the topic. This combination of drawing pictures and telling stories is a wonderful assessment tool for early childhood science education.

Evaluating Student Journals

For some purposes, you may want to use a specific scoring **rubric** or set of criteria to evaluate your students' understanding on the basis of their science journals. The following checklist is an example of such a rubric (modified from Shepardson & Britsch, 1997):

Assessing for Conceptual Understanding

1. Evidence of conceptual understanding
2. Use of information, facts, and appropriate vocabulary
3. Evidence of changes in understanding
4. Drawings represent a realistic model

Assessing for Science Processes

1. Problem stated clearly
2. Expresses procedure for investigation
3. Observations recorded
4. Variable identified (where applicable)
5. Ideas and inferences based on observations and data
6. Analyzes the investigation with possibilities for change

These categories of analysis lend themselves to various types of scoring. For example, we might allow up to 10 points for each of the 10 criteria listed above, so the maximum score would be 100. Using this system, some teachers would give each student a numerical score. Other teachers use qualitative terms for science journals—for instance, "unacceptable," "acceptable," or "excellent," depending on the number of scoring points. In younger grades, labels like "budding scientist," "scientist," and "super scientist" can be useful.

Whatever system you use to evaluate science journals, you'll find that they give you many insights into the ways in which your students construct meaning. You'll see what the children understand and how they have come to understand it. You'll also discover what, if any, alternative conceptions need to be addressed in subsequent activities. Science journals are a good example of assessment techniques that provide a window into children's thinking.

USING SCIENCE PORTFOLIOS FOR ASSESSMENT

A selection of student work

A **science portfolio** is a selection of student work produced during a semester or a unit. The work included in the portfolio may take several forms—for example, a report, a drawing, a poem, or a letter that represents what the student has learned in science over a defined period of time. What is important about the portfolio is that the student selects its contents and reflects on his or her reasons for this selection.

Portfolios are more often created in writing, art, and math than in elementary science. They would be used more often in science if more teachers understood their value. As an example of assessment activities that might form the basis for a portfolio, think back to the electricity lessons in Chapter 11. Ms. Travis might offer her students the following activities:

Sample assignments for portfolio assessment

- Write a letter to a relative explaining how a simple circuit works.

- Build a five-question game card using hidden electrical circuits. [The game card would be made with materials like aluminum foil, a manila folder, and a circuit tester consisting of a battery, three wires, a bulb, and a bulb holder. Only the card would go into the student's portfolio.]

- Describe the reason that houses should be wired with parallel circuits.

- Explain why circuit breakers are important.

- Using the letters in the word *electric,* write one statement about electricity beginning with each letter.

- Describe an activity that you have done with static electricity. What did you find out?

Each student would select and carry out three of these assignments to include in the science portfolio on electricity. As a guide to the portfolio, the students would explain the reasons for their choices. Of course, they could include any additional materials they chose. With an assessment technique like this, students become actively engaged in representing their understanding to the teacher.

An Assessment Assignment for a Unit on Electricity

As part of a portfolio assessment of what they have learned about electricity, students might be asked to write one sentence about electricity beginning with each letter in the word electric. *Here is what one student wrote:*

E   is for electrons running though a wire.

L   is for the loop in the wire when you connect it to the battery holder.

E   is for electricity that makes my hair dryer work.

C   is for a circuit, which is a pathway for electricity.

T   is for the tester that we make with a circuit.

R   is for the resistance of the load in the circuit.

I   is the symbol for amperes, a measure of electricity.

C   is for current when you measure how much electricity flows.

Guide and cover sheets

Some teachers use ordinary folders in which a guide sheet has been stapled. These folders act as the students' portfolios, and the guide sheet includes a list of the contents under headings like these:

Description of Selection   *Why I Included This Item in My Science Portfolio*

Accompanying each selection is a cover sheet that may contain the following statements to complete:

Doing this assignment helped me:

_____

My favorite part of this assignment was:

_____

What I learned from this assignment was:

_____

Evaluating Student Portfolios

Collaborating on a rating system

As with science journals, scoring systems for portfolios vary from teacher to teacher. Some teachers assign points; others use qualitative ratings.

In the higher elementary grades, the portfolio rating system is usually developed through collaborative efforts by the class and the teacher. The teacher facilitates a discussion about what the class should look for in a science portfolio. Here is one example of a guide that a class and their teacher agreed to adopt:

Guide for Evaluating My Science Portfolio

■ It contains all required items.

■ It demonstrates my understanding of the science ideas behind the unit.

■ It contains my reflections about the science experiences.

■ It shows my ability to use science process skills.

USING SCIENCE CONVERSATIONS FOR ASSESSMENT

Not all students can demonstrate what they know as well in writing as they can in "telling." An oral **science conversation,** sometimes called a **science interview**—that is, a direct communication between an individual student and the teacher—can overcome the problems that are created when we assess students solely on what they can explain in writing.

Personalized communication

Chapter 12 stressed the importance of how teachers question their students. But the usual questioning process, even something as simple as, "What do you think about this?" or "Why do you think this happened?" may leave out some members of the class. In personalized communication, you can engage a particular student one-on-one, find out what he or she is thinking, and negotiate the meanings of any terms that are problematic.

A sample science conversation

Consider the story at the beginning of this chapter about my visit to southern Texas. After the student and I straightened out the pronunciation difficulty and we both understood that we were talking about science rather than street signs, we had the following conversation:

ME: What do you do when you do science?

STUDENT: We do experiments.

ME: What is an experiment like?

STUDENT: Well, one time we measured how much popcorn we had in a cup. Then we popped the popcorn and measured it again.

ME: How did you measure it?

STUDENT: We filled up other plastic cups with the popcorn and counted how many cups we had.

ME: What did you find out?

STUDENT: We got over 20 cups of popcorn from 1 cup!

ME: Wow, that's a lot of popcorn.

STUDENT: Yeah, the whole class ate it.

ME: Why do you think there was so much popcorn after you popped it?

STUDENT: Not sure. . . I think it was something about how big each kernel got when it popped.

You can see that teachers can ask students to elaborate on statements in a science conversation, and in this way they can determine the depth of a student's understanding. Oral interviews such as these have proved to be an effective way for both teachers and students to communicate what is known (Dana, Lorsbach, Hook, & Briscoe, 1991). Teachers gain important clues about how students construct personal meaning, and these clues then influence their future instruction.

USING TECHNOLOGY TO ASSESS UNDERSTANDING

Technology-rich environments provide students with many additional ways to express their understanding to others. For example, science journals kept on a computer can incorporate images as well as words. And that is just the tip of the iceberg, so to speak, for technology use in assessment. Various presentation formats available in multimedia software programs offer new and spectacular avenues of expression for students, particularly those who may have difficulty expressing themselves in typical oral or written forms or in traditional test formats. These presentations are also a good way to challenge gifted and creative students to expand their horizons.

Many presentation formats available

Multimedia presentations of understanding typically involve text, graphics, videos, and sound controlled by a computer. Often the software

Hypermedia authoring tools

packages used for such activities are called **hypermedia authoring tools.** The term *hypermedia* refers to multimedia applications that a student can use in a flexible and nonlinear fashion, moving from one information source to several others, always in control of which options to take (Grabe & Grabe, 1998). *Authoring* means that the student can use the package to design and create his or her own product.

Hypermedia authoring environments include HyperStudio and HyperCard. Presentation software like PowerPoint can also be used for slide-show displays. (See the Resources for Further Exploration at the end of this chapter for web sites where you can examine these software products.) The combination of text, graphics, and sound is what makes hypermedia environments so versatile. The student can design multiple ways of demonstrating understanding of a science concept.

Combining text, graphics, and sound

Example of a hypermedia presentation

For example, in our lesson on land snails, a student might use HyperStudio to create a computer screen display that includes a drawing of a land snail and a short, descriptive passage of text. When the viewer clicks with the mouse on an on-screen button, the snail appears to move slowly across the screen, accompanied by a recording of the student's voice offering observations about the speed of snails. Creating such a personal design package and presenting it to the class is an excellent way of using computer technology for assessment.

As with all other types of nontraditional performance-based tasks, hypermedia design efforts must be evaluated according to an assessment checklist and rubric. If the students in Ms. Travis's fifth-grade class were asked to work in groups to design a hypermedia package that would compare series and parallel circuits, the following assessment checklist might guide them:

Sample assessment checklist for hypermedia

- Does your presentation state the goal of the program?
- Does your presentation include the materials that are part of electric circuits?
- Does your presentation show drawings of the two types of circuits?
- Does your presentation offer real-life examples of where both types of circuits are used?
- Does your presentation analyze the strengths of both types of circuits?

Regardless of the form of your students' presentations, the students themselves must be aware of the criteria for success if you are using their performances to measure the level of their understanding.

MULTIPLE TYPES OF PERFORMANCES ....

Science journals, portfolios, and hypermedia presentations are all useful methods for performance assessment. That is, they all can engage students in carrying out an assessment task that is relevant to the instructional context. These methods are far from the only possibilities, however. The following stories describe different ways in which students exhibit their understanding through performance.

SCIENCE STORY

Third Graders Enact the Water Cycle

In Ms. Nelson's third-grade classroom, the students have been exploring the water cycle and weather. On several days they have gone outside to observe the sky and make drawings of different clouds. They have been heating water and watching what happens. One day, Ms. Nelson places a pan of ice cubes over a pot with steam rising from it, and the students observe water droplets forming on the bottom of the pan. As these water droplets grow in size, they become so heavy that they fall back into the pot.

The students' experiences

During the students' investigations of what happens to water when it is heated, what happens to steam when it is cooled, and the connections between their classroom models and the weather outside, many "big ideas" have emerged:

Ideas that have emerged

When water is heated, it changes into steam.

You can see the steam rising from the pot. It looks like smoke.

When steam is cooled, it turns back to water.

The water falls back to the pot, filling it up again.

The water becomes steam again when it is heated.

When I see my breath in the winter, it is like a cloud.

Steam is really a gas called water vapor.

When students create class presentations, they express what they have learned about a particular topic. Students use many types of tools to create their presentations. This is one form of elementary science assessment.

Lawrence Migdale

Toward the end of a week of exploration, Ms. Nelson introduces three terms: *evaporation, condensation,* and *precipitation.* She asks the students to think about how these terms describe some of the big ideas they have gathered.

The students have heard the term *evaporation* often. They know that the word applies to what happened when they washed the blackboard and the water seemed to disappear. Also, they have heard the term *precipitation* in weather reports, so they can connect it with their water droplets' falling back into the pot. But they are not sure about the term *condensation.*

At this point, Ms. Nelson draws a chart with arrows depicting water evaporating, becoming part of the clouds, and then precipitating back down to the earth. The students reason that condensation must be the

Pondering new terminology

process of the gas's turning back into the water droplets. Ms. Nelson is impressed with their reasoning.

After some time, Ms. Nelson asks, "How shall we express what we know about the water cycle and weather?" The students come up with a variety of ideas:

Make a mural   Write a song

Write a poem   Write a story

Do a dance

The students' own ideas for assessment projects

Ms. Nelson asks, "What should all of our water cycle and weather projects have in common?" The students decide that their projects have to explain their understanding of *all* the big ideas they have listed that week about the water cycle.

Now Ms. Nelson guides the students in forming groups that reflect their various talents and interests. Class time is set aside for group meetings, and resources are made available. The dance group, for example, needs a tape recorder; the art group needs brown butcher paper, pencils, and colored markers.

Forming talent groups

Eventually all the groups present their projects to the class. The song group sings a song they have composed about the water cycle, with instrumental accompaniment by one of the members who has brought in his guitar. The members of the dance group, except for the student playing the role of the sun, mime raindrops in the water cycle. They start out crouched on the floor like water particles, reach up to the sky like gas particles, shiver and huddle together to form a cloud, and gracefully fall down to the ground as rain—all to the background music of "Raindrops Keep Falling on My Head." Some students have opted to write personal water cycle stories, which they read to the class. One student has imagined that she is a drop in the water cycle, and she describes her journey through evaporation, condensation, and precipitation.

The dance group's interpretation

In Ms. Nelson's class, all of the students have become engaged in the experience, thought deeply about it, made personal choices, and reflected a high level of understanding about the water cycle. The multiple assessment modes have allowed them to use various talents and means of presentation to express what they know. This is one way to implement what Brooks and Brooks (1993) refer to as "assessment in service to the learner."

Students' deep engagement

SCIENCE STORY

Second Graders Do a Station Assessment for a Unit on Matter

Setting up work stations around the room, similar to those described in the science circus in Chapter 5, is one way to design unit assessments. This format is often called a **station assessment.** The students use an answer sheet to respond to the questions at each assessment station. You must be sure that these work stations create a genuine context for the students to perform a task. Each station must challenge them to create meaning based on their prior knowledge and the experiences in which they were engaged during their science study.

Stations that present challenging tasks

Consider this station assessment developed by second-grade teachers for a science unit on matter. Their school is in the Northeast, and it has been a snowy winter. The children will visit three stations.

Station assessment for a "matter" unit

At Station 1 they find a plastic bag of air, a plastic bag of water, and a plastic bag of ice. The children are encouraged to examine these bags, but not to open them. The card on the table reads:

Station 1: Comparing, contrasting, describing

Look at the plastic bags of ice cubes, air, and water. In what ways are they the same? In what ways are they different?

At Station 2 are two identical glasses of water. One is covered, the other uncovered. The card reads:

Station 2: Predicting

> What do you think will happen to the water in the glass that has no cover on it if we leave it out on a sunny day?

At Station 3, the students discover sheets of paper for them to take. Each sheet has three boxed pictures on it. The pictures represent different stages of a snowman melting. The card reads:

Station 3: Sequencing, describing

> Take one sheet and cut the pictures out. Paste the pictures in order. What is happening in each picture?

When the second-grade teachers designed these stations, they knew they also needed an assessment checklist and rubric, so they developed the one shown in Figure 15.4. After using the assessment in their classes, all of the teachers felt it had worked well. For the students, it was an important experience. For the teachers, the class tallies of performance led to a useful evaluation of the science lessons. The results of this assessment helped the teachers modify the curriculum and instruction to meet the students' needs.

Developing an assessment rubric for the stations

FIGURE 15.3
· ·

A Paper-and-Pencil Test that Invites Students to Use Process Skills

Look carefully at these students.

1   2   3   4

4. Which of the following statements about these students is correct?
    A. Students 1, 2 and 3 all have long hair
    B. Students 2, 3 and 4 all have long pants
    C. Students 1, 2 and 4 are all smiling
    * D. All of the answers A, B and C are correct

5. A. One student has short hair
    B. One student is wearing a dress
    C. One student is not smiling
    * D. All of the answers, A, B and C are correct

From *Journal of Research in Science Teaching,* 27(8), p. 727–738. Excerpted with permission from the *Science Process Assessment for Fourth Grade Students,* © 1986 Kathleen A. Smith. Revised 1995. Reprinted by permission of John Wiley & Sons, Inc.

To develop station assessments, you need to do the following:

1. State the goal of the assessment.

2. Describe the format of the assessment tasks, making sure that they match the instructional activities.

3. Develop a scoring rubric for measuring student competence.

4. Evaluate the students' responses to the experience.

Steps in creating a station assessment

FIGURE 15.3
continued
· ·

TANK 1

Guppy swimming -- Student drops an
alka-seltzer tablet in the tank. The
bubbles are carbon dioxide.

TANK 1

After one minute, guppy
stopped swimming and had
trouble breathing.

TANK 2

Guppy swimming -- Plain water.

TANK 2

Guppy swimming -- Plain water
<u>after</u> one minute.

8. Which sentence would best describe what effect alka-seltzer has on a guppy?
   * A.   Guppies may not be able to survive long when carbon dioxide is in the water.
     B.   Guppies become active when carbon dioxide is in the water.
     C.   Guppies show <u>no</u> change in behavior when carbon dioxide is in the water.
     D.   All of the answers A, B and C are correct.

From *Journal of Research in Science Teaching,* 27(8), p. 727–738. Excerpted with permission from the
Science Process Assessment for Fourth Grade Students, © 1986 Kathleen A. Smith. Revised 1995. Reprinted
by permission of John Wiley & Sons, Inc.

ASSESSMENT AND THE NATIONAL SCIENCE EDUCATION STANDARDS ·

**ASSESSMENT AND
LEARNING ARE TWO
SIDES OF THE SAME
COIN.**

*—National Science Education
Standards*

The National Science Education Standards support the importance of creating assessments that have a real-world relevance and context. This type
of authentic assessment is the mirror image of the authentic instruction
model that the NSES promote and this book describes. If we engage students in meaningful science experiences that relate to real-world contexts,

then we need to assess them with meaningful experiences that relate to real-world contexts. To put it another way, when we allow them to be active participants in their own learning, we should assess their understanding with tasks that actively demonstrate what they know and are able to do—tasks that match the instruction.

NSES criteria for authentic assessment

The NSES state that assessment tasks are authentic when they ask students to apply their science knowledge and reasoning to "situations similar to those they will encounter in the world outside the classroom, as well as to situations that approximate how scientists do their work" (National Research Council, 1996, p. 78). Suitable activities include, but are not limited to, planning an experiment, extending an investigation, designing charts and graphs to represent data, constructing a model, and participating in an interview.

This chapter has advocated extensive and multiple means of performance assessment, including problem solving, creative writing, drawing, drama, technology applications, and other original forms of presentation.

When to use paper-and-pencil tests

But, you may ask, aren't there any traditional paper-and-pencil tests that would be useful? You can indeed use paper-and-pencil tests, with multiple-choice or fill-in questions, to assess your students' familiarity with terminology and their recall of particular bits of science knowledge. Also, paper-and-pencil assessments have been constructed to measure the science process skills of elementary school students (Smith & Welliver, 1990). In the example shown in Figure 15.3, the test questions were matched to specific science process skills. Looking at the figure, you should be able to identify questions that relate to the skills of observation, inference, comparing, contrasting, and so on. But notice that although each question engages the student in using a process skill, there is no apparent reason to do so. There is no real-world context. Hence, tests of this type may have limited usefulness for teachers who want an active demonstration of their students' understanding.

Limits of these tests

The NSES do include paper-and-pencil tests as one assessment option. In stressing authentic assessment, however, they point us toward more active, performance-based assessment strategies:

For instance, a student's ability to obtain and evaluate scientific information might be measured using a short-answer test to identify the sources of high-quality scientific information about toxic waste. An alternative and more authentic method is to ask the student to locate such information and develop an annotated bibliography and a judgment about the scientific quality of the information. (National Research Council, 1996, p. 84)

FIGURE 15.4   **Sample Checklist and Scoring Rubric for a Station Assessment**

The following hands-on assessment of the second-grade unit on matter will provide us with a better understanding of what children know and are able to do as a result of their experiences with the matter unit.

1. Stations 1 and 2 invite observation, comparing, classifying, and describing.
2. Station 2 asks children to predict an outcome they cannot observe at the moment.
3. Station 3 asks children to place events in their proper time sequence and describe the event in each picture.

Scoring Rubric—Matter Assessment

<u>YES</u>   <u>NO</u>

1. Uses scientific terms such as:
   solid, liquid, gas, evaporation.
2. Accurately describes the water.
3. Accurately describes the ice.
4. Accurately describes the air.
5. Makes reasonable predictions about the
   glass of water that is uncovered:
   Water will get warm.
   Water will disappear.
6. Correctly sequences events.
7. Describes pictures accurately.

The following categories reflect the number of "yes" checks:

Budding scientist	Scientist	Super scientist
2–3	4–6	all 7

Reprinted by permission.

TEACHING, LEARNING, AND ASSESSING

Carefully designing the assessment
Since educational data profoundly influence the lives of students, the design of assessments requires careful consideration. The drive to support authentic instruction with authentic assessment requires that teachers

state the purpose and design of the assessment, as well as the ideas and processes it is intended to measure (National Research Council, 1996, pp. 78–79).

As you have seen throughout this chapter, a good assessment task fits into the science experience and is a good instructional task as well. The students do not feel judged or threatened, and the teacher discovers what they know and understand about a science idea. Understanding what your students know about a topic will then influence your science instruction. Hence, assessment and instruction are interactive.

Assessment in a diverse environment

Remember that your class will probably include diverse groups of students with diverse learning styles and multiple ways of communicating their ideas. If you vary the types of assessments you employ, you create a more equitable learning environment. The chances are that with multiple modes of assessment, you will find a match for the mode of expression with which each student feels most comfortable. When you use a single assessment approach, you place students who are not comfortable with that selected way of expressing themselves—or children who have disabilities in a specific mode of communication—at an unfair disadvantage. In designing an assessment, make it part of your whole continuum of teaching and learning experiences. Keep asking yourself questions like these:

Questions to ask yourself about assessments

- Who are my students?

- How can I learn about what they understand?

- In what ways do they express themselves?

- How are the science experiences I facilitate helping them to construct meaning?

An ongoing process

Assessment is ongoing. It is embedded in all the work you do with children in elementary science. The children are assessing themselves through their activities. You assess them as you visit their groups. They keep records of their experiences. You engage in conversations about science ideas with the students. Scientific inquiry is a complex process, and we need many types of activities to assess what the children know and are able to do.

We are hoping the children will begin to understand the ideas of science and how they relate to the ideas in the world around them. As you move to the next chapter, start to think about that wider world. As Chapter 16 will show, we can see science as a community effort. We can construct a model of science that engages children, their families, businesses, and other agencies in the wider community, all contributing to the understanding that science is a special way of knowing the natural world.

Key Terms

science assessment *(p. 311)*
testing *(p. 311)*
performance assessment *(p. 311)*
authentic assessment *(p. 311)*
evaluation *(p. 312)*
science journal *(p. 313)*
rubric *(p. 320)*
science portfolio *(p. 321)*
science conversation *(p. 323)*
science interview *(p. 323)*
hypermedia authoring tools *(p. 325)*
station assessment *(p. 329)*

Resources for Further Exploration

Electronic Resources

Arter, J. A., Spandel, V., & Culham, R. (1995). *Portfolios for Assessment and Instruction.* **http://www.uncg.edu/~ericcas2/assessment/diga10.html**. An online article from the *ERIC Digest* that explains the uses of portfolios in instruction and assessment.

ERIC Clearinghouse on Assessment and Evaluation. **http://ericae.net**. A site that offers links to resources on a variety of assessment topics, including performance and portfolio assessment.

HyperStudio. **http://www.hyperstudio.com**. A web site from Roger Wagner Publishing that offers an introduction to HyperStudio software.

Sweet, D. (1993). *Student Portfolios: Classroom Uses.* **http://www.ed. gov/pubs/OR/ConsumerGuides/classuse.html.** Part of the online Consumer Guides series published by the Office of Educational Research and Improvement, this article discusses how to develop portfolios and where to find more information about them.

Welcome to HyperCard. **http://www.apple.com/hypercard**. Includes background information on HyperCard software.

Welcome to Microsoft. **http://www.microsoft.com**. From Microsoft's home page, you can follow links to descriptions of various Microsoft products, including PowerPoint.

Print Resources

Dana, T. M., Lorsbach, A. W., Hook, K., & Briscoe, C. (1991). Students showing what they know: A look at alternative assessments. In G. Kulm & S. M. Malcom (Eds.), *Science Assessment in Service to Reform*. Washington, DC: American Association for the Advancement of Science.

Hein, G., & Price, S. (1994). *Active Assessment for Active Science*. Portsmouth, NH: Heinemann.

Moline, S. (1995). *I See What You Mean: Children at Work with Visual Information*. York, ME: Stenhouse.

Shepardson, D. P., & Britsch, S. (1997). Children's science journals: Tools for teaching, learning, and assessing. *Science and Children* 5 (34):13–17, 46–47.

16 Involving the Community in School Science

FOCUSING QUESTIONS

- Do you ever hear adults say, "I was never very good at science"?
- What are some sites in your community for informal science education?
- Do you know any adults who are research scientists by profession?

When I was in third grade, I made an anemometer and placed it on the fire escape of the apartment building where I grew up. An anemometer is a weather instrument that can measure wind speed. My parents helped me build it with pieces of lumber and large cups to catch the wind (see Figure 16.1). It lasted until the first rainstorm. What interested all of us about the anemometer was that we could tell how windy it was outside without opening the window. I chose to make the device for a school project, and my parents helped. But this was the only time I remember a science-related topic from school that engaged my entire family.

The story of my third-grade anemometer should remind us that parents and other significant adults are our students' first teachers. Engaging these adults in science experiences with students is a way not only to bring families together, but also to encourage a deeper interest in science and science-related activities. Research tells us, moreover, that parents' attitudes affect their children's attitudes, and some adults' fears of science have a negative impact on their children (Kober, 1993). By emphasizing the relevance of science *outside* school—in our students' everyday lives—we can help change those negative images of science. In this way, we can promote science both inside and outside the classroom.

In this chapter, we explore some ways to involve families and other community members in the process of doing science with children. We will look at a family science night, a science open house, an interview with a scientist, and visits to community science resources. Keep in mind that these are only a few of the possibilities for community involvement.

PRIMARY SCHOOL FAMILY SCIENCE NIGHT

A skeleton greeted the visitors to a local school. Hanging from it, a sign read, "Welcome to Family Science Night." This elementary school for the primary grades was hosting a family and community event designed to illustrate the joys and challenges of doing science. At the same time, the event demonstrated to the community that, at this school, "doing science" was central to the children's education.

The family science journal

A **family science night** can take many forms. This one was spread over two nights so that as many families as possible would have the opportunity to participate. As parents and children filed into the school cafeteria, they picked up red booklets titled, "My Family Science Journal." All around the cafeteria, signs offered sayings like, "Science is not about knowing but about trying to find out." Next to the red booklets were drawing paper and markers, and the children were invited to "draw a scientist" as an introductory activity.

What does a scientist look like?

When the excitement of gathering together had passed, the parents and children sat in arranged seats for a few moments as the chair of the event, Ms. Amdur, greeted everyone and asked the children to come up front to display their pictures of scientists. Although many of the children's pictures were of mad, male, stereotypical scientists, some children drew females and some even drew self-portraits. Ms. Amdur then showed the audience what a true scientist looked like by holding up a large mirror so that the children gazed at their own reflections. This placed the children and their ideas about science at the center of the experience. Then the school principal and the science committee welcomed the families, and everyone was on the way to a meaningful night of science.

FIGURE 16.1
· ·

A homemade
anemometer.

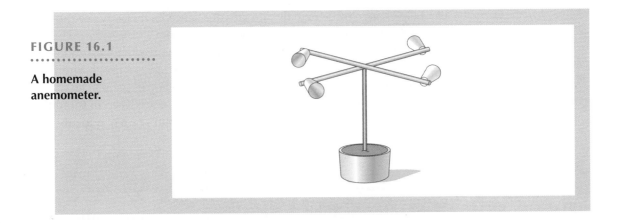

Background and Philosophy

At this school the family science program was established to create a bond between home and school, as well as to encourage interest in science and promote science achievement for both girls and boys. Family Science Night evolved after teacher representatives from each grade formed a science committee and worked together for several years to develop a science curriculum for each grade level. Once the curriculum was in place, the teachers felt they were ready to extend the science activities to the children's families.

Alternative families welcome

In the flyer that invited people to participate, "family" was loosely defined, and the children were asked simply to bring a grown-up. Families with alternative caretaking structures were welcome, as were any other special adults in the children's lives. The adults who came included parents, aunts and uncles, grandparents, guardians, older siblings, friends, "big brothers" and "big sisters," and others.

As a guiding philosophy for its science program, this primary school favored a hands-on, inquiry-based approach in which children constructed knowledge through their own experiences. For Family Science Night, the teachers followed the same approach. They designed seven stations that invited manipulation of materials and trial-and-error problem solving. The activities were carefully planned to stress process skills like observing, inferring, predicting, recording data, building models, identifying variables, drawing conclusions, and planning investigations.

Problem solving for families

Although they intended to highlight the process skills of science, the teachers did not want to duplicate activities that the children performed in

A Family Science Night engages parents, grandparents, and guardians in science activities. Here a father and a daughter solve a problem at one of the many science stations.

Lorraine Amdur/Somers Public Schools

the regular science curriculum. The teachers also decided that the activities should be:

- Practical to set up.
- Not overly costly.
- Suitable for maximum interaction between parents and children.
- Representative of the subject areas of environmental science, physical science, and life science.

Design of the stations

Each station was designed to engage families in several tasks. The stations were complete with directions, materials, and questions to encourage higher-order thinking and discussion. Each activity had a corresponding journal page on which the families could write or draw about what they were doing. The children were encouraged to act as the journal writers for the family. The journal also included questions to guide the students and follow-up activities to perform at home.

Let's look at the seven different stations that the children and their family members encountered.

The Science Stations

How many drops of water on a penny?

Penny Station   At this station, the families explored how many drops of water could fit on the head of a penny. Then they explored the water droplet as a magnifier. They were asked to record their data on the corresponding science journal page.

Canal Races   This investigation used "boats," made of pint-size milk cartons, string, and washer weights, that raced down "canals" made of wallpaper bent into troughs. The families were asked to compare the speeds of two boats attached to different-sized masses.

How to make the boat go faster?

Each family assembled its own boat by attaching weights to the end of a premeasured length of string and then attaching the string to a milk carton. The string was exactly the length of the wallpaper troughs (Barrow, Knipping, & Lutherland, 1996). After laying the string in a wallpaper trough with the weight hanging off the back edge, the family would drop its boat gently over the front edge of the trough. Pulled by the weight, the boat would race along this homemade canal. Families explored how to make their boat go faster by adding weights. Prior to the activity, the staff had assembled a "pace boat" that families could use for comparison.

How does exercise change heart rate?

Heart Rate Center   During this activity, the children listened to their heartbeats with a stethoscope. They listened twice, before and then after a strenuous exercise, and counted the number of beats in a fixed period of time. They recorded these before-and-after heart rates in their science journals.

What keeps a bear warm?

Polar Bear Model   This investigation addressed how the polar bear stays warm in icy water. Families immersed their hands in icy water twice: once wearing fleece mittens and then with fleece mittens covered with lard-filled plastic bags. They found that the lard, a model of the polar bear's protective blubber, offered better insulation. Families were amazed at how warm their "polar-bear hands" were. Figure 16.2a shows the questions they answered in their science journals.

How does a snail move?

Snail's Trail   For this investigation, the families gently handled and explored the movements of garden land snails, noticing the secretions of the snails as they moved along colored paper.

FIGURE 16.2
....

Family science journal
pages from a primary
school's Family
Science Night.

POLAR BEAR

Which hand stayed warm the longest?

Why did that hand stay warmer?

How does our MODEL tell us what keeps a polar bear warm?

At home: Create a model to show how an animal in our area adapts to its environment.

A

THE OIL SPILL

What was the best way to get the oil out of the water?

What is the best way to get rid of the oil-soaked materials you now have?

How would a bird clean its feathers after an oil spill?

At home: Find signs of pollution in your area. Plan a way to eliminate the pollution.

B

Reprinted by permission.

How can you get oil out of water?

Model Oil Spill   One environmental science activity addressed cleaning up a model oil spill in a large foil baking pan. The families had string, cheesecloth, paper towels, cornstarch, and sponges to work with. They found the task was not easy, and they created another mess in the process! See Figure 16.2b for the questions that the families answered in their science journals.

How can you build an earthquake-proof structure?

Stable Structures   This investigation explored how structures could withstand an earthquake, modeled by fists pounding on the table. Families attempted to build stable structures from 50 wooden cubes, each 2.5 centimeters on a side. Then they explored how their structures withstood their mock earthquake.

Building a Tradition

The science journals contributed a great deal to the success of the evening. Families used their journals as a way to keep track of what they did and what they learned. Further, the journal became a symbol of science and of Family Science Night. At home, families could use their science journals to duplicate the science night activities.

An annual tradition — with community input

In this school, Family Science Night became an annual tradition. The teachers changed the station design each year. Letters went home with the children asking community members what they would like to see at the next family science event, establishing an ongoing science link with the students' homes.

This story is just one example of a successful format for a community science night. In the next section, you will see the impact of another sort of community interaction: bringing a local scientist to visit a group of students.

A SCIENCE CONVERSATION—WITH A REAL SCIENTIST ....

Rosario Enriquez was a physicist at a research laboratory in the community where I was teaching. I met her because we were both active in the local chapter of the Association for Women in Science, an organization that promotes women's careers in science. After growing up in Mexico, Rosario had come to the United States to complete her education in physics. She was interested in designing devices to assist people with artificial limbs.

I invited her to meet the elementary school students who visited my science room in the local school. I told her I felt it was important for children to meet practicing scientists, and she agreed.

The students prepare

The day before she came to the science room, I explained to the students that we were going to meet a scientist and have the opportunity to conduct an interview. They had lots of potential questions prepared. I did not mention the gender or ethnicity of the visitor we were expecting.

The first encounter

Rosario was chatting with me in the science room when a third-grade class entered. Shortly after, a boy approached me and asked when the scientist was coming. "She's here," I said, and I introduced Rosario to the group. Clearly Rosario did not match this boy's image of a research scientist. In fact, many of these students had had little or no contact with real scientists, and their culturally constructed image of a scientist was based, in part, on media portrayals—the stereotypical scientist as a man in a white lab coat.

The scientist's story

The questions that the children had developed the day before addressed Rosario's school science experiences, her decision to become a scientist, memories of special science teachers, and the nature of her work and her workdays. Rosario described her life in Mexico. Her father was a scientist and taught at the local university. She had always known she would be a scientist, and she talked about her love of science and scientific

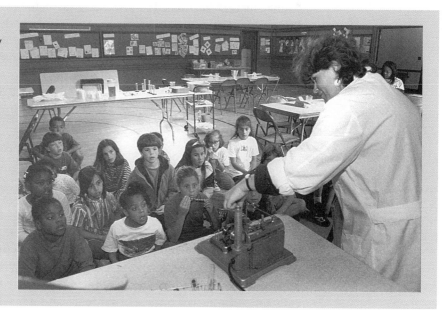

Sometimes community scientists visit elementary students and set up demonstrations in the school. This scientist is introducing second and third graders to the principles of evaporation.

Elizabeth Crews/The Image Works

research. She explained to the children that science "is not about finding the answers as much as it is about asking the right questions." Rosario also explained that she had decided to do laboratory research on artificial limbs because she was interested in using scientific study for the good of the people.

The children acquired a better understanding of science as a human enterprise through the interview. They came to appreciate that what keeps scientists going is the fascination of the search and the "re-search." The experience brought home the centrality of science as process, as an activity involving searching and problem solving. The students realized that scientists are frequently repeating experimental work, that science is not neat and tidy, and that science requires patience and perseverance. They also learned that scientists are real people willing to share their time and their work with other members of their community.

Learning about the search and the "re-search"

Further, the children learned an excellent cultural lesson. When they saw a young-looking Mexican woman in their science lab, they did not at first associate her with their visiting scientist. This was an important message of Rosario's visit. By inviting scientists and engineers from varied backgrounds to visit our students, we can deconstruct some of the cultural stereotypes surrounding science.

Challenging a stereotype

To set up a similar interview in your own class, you can look for a research scientist in your community or engage in electronic dialogues with scientists through email. Many scientific institutions make a point of sponsoring such interactions. For instance, Massachusetts General Hospital in Boston has a science mentor program that sends doctors into schools to visit with children. Through such institutions and through personal contacts, the chances are that you can locate a scientist who has some commitment to the community or even to your particular school. This person will bring a real-life perspective to the science work in your classroom. He or she will let the children know that in order to *learn* science, you have to *do* science.

SCIENCE OPEN HOUSE

A kaleidoscope of activities

A **science open house** involves a demonstration by the students to the community of one or more science experiments, activities, or models. This community event, usually for older elementary grades (grades 4–6), typically involves several groups of students, and it is set up in different classrooms in the evening. In fact, the science open house can be a kaleidoscope of science activities, each attracting a group of parents and other community members to different spots in the school.

The flyer announcing the open house can alert the community to the types of science activities going on in each room. There may be as many as ten groups of students doing science with the community members. An open house often engages the adults in simple science activities in which the children have been engaged in their regular classrooms. For example, student groups might set up investigations on topics like these:

Sample topics for open-house activities

Batteries, bulbs, and wires

Different types of liquids

An edible model of the solar system

Using cabbage juice to test the acidity of household liquids

Measurement with the metric system

Making models of the organs of digestion

The science open house is a way to give families and the community a close look at the school's science program, demonstrating science as both a process and a set of ideas.

BEYOND THE CLASSROOM: COMMUNITY SCIENCE RESOURCES ....

As we discussed in Chapter 6, taking students out into the community to do science—on field trips and other excursions—is an important way of exploring the natural world. Some communities offer many opportunities to explore science and science-related topics, including opportunities that we do not necessarily "code" as relating to science or nature.

Visiting local utility plants

For instance, in addition to exploring the natural fauna and flora of your surroundings, you may want to investigate how your community recycles waste or produces electricity. Students are often amazed by a trip to the local electric power plant, and their electricity unit is enhanced by this local connection that affects their everyday lives. Similarly, with a trip to a waste-to-energy plant, you can explore the burning of garbage to produce electricity. Other manufacturing facilities may offer similar connections to your science curriculum.

Other real-life connections

If your community is rich in farmland, the treatment of soil and sources of water are rich real-life connections for any unit on plants. Some vertebrate units in elementary school can be enlivened by a visit to a local fish hatchery. In some communities, the invertebrate unit can be supported by a visit to a local fossil site, where ancient marine invertebrates have left their mark on the now-uncovered rock.

Urban communities offer other types of enrichment, such as museums and library events. Some community museums sponsor sleep-over events in which youngsters spend a day and a night engaging in science activities. Some museums collaborate with other community organizations, planning events together. For example, at the Franklin Institute in Philadelphia, two projects combine community outreach with museum-based activities. These collaborations specifically target families from groups identified as underrepresented in science (McCreedy, Borun, Mostache, & Wagner, 1996). One of these projects, called Girls at the Center, builds on existing partnerships between science museums and Girl Scout Councils. The museum's role is to provide planned, inquiry-based science learning opportunities for the girls and their families. The Girls at the Center project also has sites in Arizona, Florida, Indiana, and Minnesota. The other such collaboration at the Franklin Institute, the Community Connections Project, involves four Philadelphia-area museums and eight community-based organizations. This project uses family-based programs to introduce African American and Latino groups to science museums.

Many other museums around the country work to promote family participation in science experiences, creating programs that are responsive to both adult learners and children. These programs tend to work best when schools and community-based institutions collaborate, addressing together the competition that the programs face from family obligations like work, day care, and regular school studies. When local institutions work together—combining museum-based efforts, school-based efforts, and community group efforts—the partnership can have considerable impact. These partnerships reflect the important realization that scientific skills, critical thinking, and scientific knowledge are necessary tools for everyone.

Museum outreach programs

Planning for Local Outreach

You may be thinking, "Oh, my! Creating curriculum, inviting inquiry, *and* reaching out too? How can I do all that?" Although the task may seem overwhelming at first, you will see, as you gain experience, that the science climate you and your students create in your classroom will encourage you to seek connections well beyond the school. You can become an all-important link between your students and the science resources in the wider community.

The following hints will help you in your work to reach out:

- Think about your science units. Where will local connections work best?
- Visit the local chamber of commerce, and get descriptions of local resources.
- Put your name on the mailing lists of local museums and science-related institutions.
- Visit some of these institutions, introduce yourself, and make personal contacts. Explain that you are interested in forging connections with the institution on behalf of your students' science learning.
- Look for bird-watching groups, local arboretums, and bird sanctuaries. Contact the directors or leaders, and ask for information.
- Contact community groups, and speak to their public relations personnel.
- Learn what, if any, community-school-museum collaborations exist.
- Search the *Science Adventures* site on the Internet that was listed in the Resources for Further Exploration in Chapter 6 (**http://www.science adventures.org**). Look there for local science events and institutions.
- When you do arrange a community-based activity for your students, always make connections between that experience and their in-class work.
- Ask students to talk and write about their experiences beyond the classroom.
- Create portfolios, folders, and board displays with titles like "Science in Our Community."

Using community resources helps to change students' notion of science. Rather than seeing it mainly as a subject learned in school, they come to appreciate science as a way of exploring the natural world and understanding the problems that face all of us in our interactions with the natural environment. Community events also contribute to students' learning by giving them opportunities to engage in science activities with the adults in their world.

Key Terms

family science night (*p. 339*)
science open house (*p. 346*)

Resources for Further Exploration

Organizations

American Zoo and Aquarium Association, 7970-D Old Georgetown Road, Bethesda, MD 20814. Ask about the school outreach program for protecting wildlife.

Association of Science and Technology Centers (ASTC), 1025 Vermont Avenue, NW, Suite 500, Washington, DC 20005. Ask about the Youth Alive Program for adolescents—a program that offers multiple out-of-school science-related experiences and activities.

4-H Series Project Office, University of California at Davis, HCD, Davis, CA 95616-8523. Ask about YES (Youth Experiences in Science), an after-school science activities program.

National Geographic Society, 1145 17th Street, NW, Washington, DC 20036. Ask about the *Kids Network,* a computer- and telecommunications-based science curriculum using local environmental concerns.

National Science Teachers Association, 1840 Wilson Boulevard, Arlington, VA 22201-3000. Ask for a copy of the October 1996 issue of *Science and Children,* Vol. 34(2). This issue is devoted to bringing families and science together.

Electronic Resources

Paulu, N., with Martin, M. *Helping Your Child Learn Science.* **http://www.ed.gov/pubs/parents/Science.** Sponsored by the Office of Educational Research and Improvement, this online book offers a number of science activities for parents to undertake with their children, at home and in the community. Besides recommending this resource to parents, teachers themselves can benefit from the ideas presented here.

Science Adventures. **http://www.scienceadventures.org**. A web site with information about zoos, parks, planetariums, and other community resources, searchable by zip code.

Science Learning Network. **http://www.sln.org**. A site with links to museum web sites and similar resources throughout the country.

Print Resources

Barrow, L., Knipping, N., & Litherland, R. (1996). Evening science: Solving science problems. *Science and Children,* 34(2), pp. 20–23.

Gardner, D. H. (1996). Bringing families and science together. *Science and Children,* 34(2):14–16.

Kober, N. (1993). *Edtalk: What We Know About Science Teaching and Learning.* Washington, DC: Council for Educational Development and Research.

Koch, J., Amdur, L., Ciota, M., & Johnston, I. (1998). A primary school family science night. *Science and Children,* in press.

McCreedy, D., Borun, M., Mostache, H., & Wagner, K. (1996). A collaboration for education. *Science and Children,* 34(2): 38–41.

17 Pulling It All Together: Reflection and Self-Assessment

FOCUSING QUESTIONS

- Have you started a science journal?
- Have you been exploring science web sites?
- Have you examined one or more science lessons?
- Have you located your scientific self?

When you set up a plan for doing science with your students, you need a way to pull the lesson together. This chapter is my way of pulling this book's ideas together for you as you continue to develop as a teacher of elementary school science.

We close *Science Stories* where we began Part One—with our own scientific selves. You may wonder why I return so regularly to our scientific selves. It is because we do our best science teaching when we *model inquiry* to students.

YOUR SCIENTIFIC SELF

You would like, I'm sure, to show your students what it looks like to wonder, to be curious, to get messy, to explore nature, to yearn to find out about the natural world, and to be thrilled at the prospect of experimenting. In order to convey these feelings to our students, we need to have them authentically inside ourselves.

My definitions of science change. Do yours?

My own feelings about science are not static. My definitions of science change as I grow and develop, and I am always adding to my list. I invite you to develop your own definitions of science as you have opportunities to do science with children. The following list of definitions contains statements that are true for me at different times. I usually do not believe all these definitions at the same time. They change as I change, as I interact with nature and with children.

Some of the ways I define science

- Science is a way of exploring nature.
- Science is both a method and a set of ideas.
- Science is a subject in which you expect the unexpected.
- Science is an area of study that can be frustrating (like when the land snails do not arrive on time).
- Science is a way to explore students' thinking.
- Science is a dynamic area of study, ever changing, always revealing new evidence, and its practitioners are always changing their minds.
- Science is learning what your students think.
- Science is a joyous activity when shared with children.

Which of these definitions do you agree with now? Have your definitions changed as you read this book? What other definitions can you add?

Getting comfortable with your scientific self

Your scientific self and your character as a science teacher are not separate from each other. A large part of your journey toward becoming a successful science teacher involves locating your scientific self and becoming comfortable with it. Find ways in which you are comfortable exploring nature, write about them in your science journal, think about the characteristics of exploration, and come up with *your own* ideas and test them. Remember that we use process skills in our daily life, so look for experimentation in everyday places. Watch for opportunities to observe, infer, predict, classify, and experiment. Some adults experiment in the kitchen or the garage or the garden. Ever curious about your environment, take these multiple opportunities to ask yourself:

LOOK FOR EXPERIMENTATION IN EVERYDAY PLACES.

—Janice Koch

Children and teachers become science learners together. Here second and third graders are examining plant roots with their teacher.

Elizabeth Crews/Stock, Boston

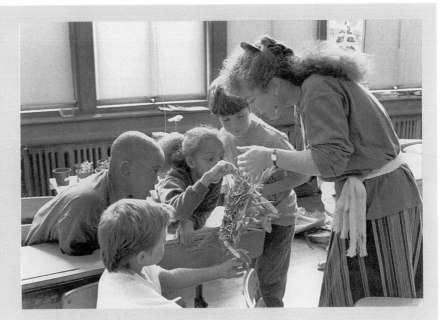

- Where is "nature"?
- What do I notice?
- What do I wonder about?

Write down your thoughts, and then share them with your students.

When you find your scientific self, you can no longer view science as just a subject taught in school. As you open your eyes to nature, you also feel a sense of confidence about exploring the scientific reasons for natural events. Look in your newspaper for items of interest that relate to scientific discovery; share these with your students, and encourage them to do the same. Ask yourself: What contemporary scientific issues especially engage me? Are they environmental issues? Do they relate to genetic engineering? Health and nutrition?

Looking around for science opportunities

Similarly, take time to visit local community resources on your own, with your friends, or with your family. Look around for zoos, museums, botanical gardens, science and technology centers, bird sanctuaries, and other informal science learning sites. Watch even for dandelions in the cracks in the sidewalk. Being alert to all these opportunities to do science will help you become a successful science teacher as well as a lifelong science learner.

Everyday Wonderings

In the course of your daily life, what do you wonder about? Do you write your questions and your discoveries in your science journal? The following excerpts from the science journals of preservice teachers illustrate how many ordinary things you can find to wonder about.

The ocean is beautiful in the Gulf of Mexico. The water is so blue; the sand is soft. There really weren't any shells. How come some beaches are just plain sand with a few shells, and other beaches have a ton of shells? Do some places have more shells to be washed on shore? Or is there a rougher tide at the beaches with shells? I am curious. Are there more shells after a hurricane or a big storm?

Every time I bake cookies, I jump as I hear the noise of the bending pans come from the oven. I've never understood why they warp like that. If heat causes things to expand, wouldn't the pan spread yet still lie flat?

Today some of my friends and I celebrated their graduation. We all got champagne with a strawberry in it. I watched this strawberry. When it was first put into the champagne, it sank to the bottom of the glass. But as the bubbles in the glass were dwindling, some of them stuck to the strawberry. The strawberry then floated to the top of the glass. I wonder what caused this.

Well, the weather is terrible today (as usual). The humidity is definitely the worst. It makes my hair look terrible. Why is that? I mean, I know what humidity is, but why does it affect my hair so much?

Reprinted by permission.

BECOMING A REFLECTIVE TEACHER

Reflection is active, not passive.

Locating your scientific self takes time, and it often requires reflecting on nature and the way nature presents itself in your world. This reflection is an *active process*. That is, it takes active, deliberative thinking, not passive musing. Further, it often leads to additional action to consolidate your understanding. When you reflect on an aspect of nature, you may be impelled to ask questions about it, to learn more about it, to make sense of it. The bird story in Chapter 2—the entries from my science journal about the birds in the tree behind my house—is an example of how reflection stimulates further learning.

Being a reflective teacher also means frequent reflections on your students' thinking and on your own teaching practices. These reflections, too, lead to further action. Reflecting on how your students are responding to a lesson may require you to alter your lesson plan and change the course of subsequent lessons. On a larger scale, being reflective about your teaching practice will prompt you to seek out additional means of professional development, such as workshops and contacts with other teachers.

Developing a Personal Philosophy

Reflecting on your own philosophy

As you begin your teaching career and engage children in scientific experiences, you will gradually become an expert on how children construct their own meaning. You will develop your own personal philosophy of teaching, based on your beliefs and the evidence you have gathered from your experience. Reflection can help you in this process. By thinking about what you firmly believe and what you remain unsure about, you can promote your development as a teacher.

A personal philosophy statement

One useful technique is to draw up your personal philosophy statement. It should contain your thoughts and feelings and indicate their connection to your actions as a teacher. As one such exercise, try completing the following statements (adapted from Kochendorfer, 1994):

I am the kind of science teacher who thinks that . . .

When I do science with children, I feel . . .

The science experiences that I believe are most worthwhile include . . .

If you can complete those sentences, you already have a rough draft of your philosophy statement. But it undoubtedly will change over time. Consider engaging in this sort of reflection during every school year, and keep track of your philosophy statements. You may be amazed at your own growth as a scientist and as a teacher.

PROFESSIONAL DEVELOPMENT

Professional development is the process by which teachers strive to improve their work as teachers, to grow in their profession. Your professional development is based, most of all, on inquiry about your own teaching—inquiry that leads to research and action. Reflection is part of that inquiry

Value of networking

Types of networking

process, but there are numerous additional activities and resources to which you can turn. Many of them involve interactions with your professional colleagues, either within your school or in more extended networks.

Networking in the Profession

As you teach science in the elementary school, you will have many opportunities to interact with colleagues and join networks of teachers whose personal philosophy about teaching and learning is similar to your own. Local teacher centers and science organizations offer workshops and courses that create such networking opportunities for you.

Why should you spend time on professional networking? Research tells us that teachers who engage in professional communication with their peers build relationships that promote their own standards and foster reflection. With regular collegial exchanges, teachers share their knowledge and refine their practice throughout their careers (Darling-Hammond, 1998).

Networking activities can include experiential learning—for example, working with colleagues on concrete tasks, such as developing a science

Teachers learn with and from each other. Formal and informal teacher networks help elementary school teachers share strategies for doing science with children.

Carol Palmer

Sometimes teachers combine classes and work together to "team teach" — the sharing of ideas with each other and their students. Here, two teachers plan a unit on animal diversity.

Bob Daemmrich/Stock, Boston

station assessment. They can also include working with a mentor teacher in your own school, observing an exemplary teacher, and attending teacher discussion groups centered on teaching practice.

Look for such opportunities in your own school and district. Be alert for chances to attend elementary science workshops. Inquire whether your school or district will offer funding for you to attend national or regional conferences.

Joining NSTA

As a first step, you can join your state science teachers' association and the National Science Teachers Association (NSTA). Membership in the NSTA includes a subscription to *Science and Children,* a journal devoted to preschool through middle-level science teaching. The address is listed in the Resources for Chapter 16.

Promotion and Certification Requirements

Requirements vary

Increasingly, teachers are being required to show evidence of professional development in order to earn promotions, or even to have their teaching certificates renewed. Since requirements vary from state to state, you should explore the criteria in your own locale. Ability to use educational technology is a particular concern in many state certification standards, and some states have other special conditions related to science teaching.

The NBPTS

For experienced teachers, the National Board for Professional Teaching Standards (NBPTS), established in 1987, is developing a series of assessments that will certify teachers in certain fields. For elementary school teachers, there are generalist certification standards for teaching students in early childhood (ages 3–8) and middle childhood (ages 7–12). Specific standards for science teaching in middle childhood are under consideration.

Advocates of the national board believe that offering professional credentials will bring prestige to the teaching profession, and they hope that schools and school districts will see NBPTS certification as an important teaching qualification. You may want to visit the NBPTS web site, listed in the Resources at the end of this chapter, and explore the requirements for certification. You must have been teaching for a minimum of three years and have completed your baccalaureate degree before being eligible to take the series of assessments that lead to an NBPTS certificate.

Further Means of Professional Development

Besides networking with colleagues and fulfilling state and local requirements for advancement in the field, there are many other ways that you can promote your own professional development as a science teacher. Here are some ideas:

Doing your own
science research

■ Do research on science topics. If there is a topic that you wish you knew more about, turn to the library, the Internet, and science trade books to research it. Be sure to gain confidence in the science content areas that you will be exploring with your students.

Consulting curriculum
guides

■ For exemplary science instructional materials, refer to the curriculum guides listed in the Resources for Further Exploration of Chapter 14, as well as the other resources mentioned throughout this book. Also use Chapter 14's checklist for evaluating science curricula (page 308).

Trusting your judgment

■ Trust your own judgment. If you come across a scientific investigation that does not appeal to you, most likely it won't appeal to your students either.

Keeping your science journal

■ Keep a science journal, for both in-class and outside-of-class science activities.

Doing your own research on teaching

■ Do your own research on what is happening in your classroom. The term **action research** refers to research projects in which classroom teachers explore some area of their teaching or some aspect of the students' learning. For example, they may study how their students are experiencing a science unit, lesson, or activity. The goal of action research, sometimes called *teacher research* or *classroom research,* is to improve your own or a colleague's teaching. Sometimes an action research project is a questionnaire you distribute to students to learn what they thought about an investigation or a unit. Sometimes the research design is more extended, depending on what you want to find out. Doing research in your own classroom is not complicated, and the Resources at the end of the chapter can help you get started.

Writing an article

■ Write your own article about science learning for *Science and Children* magazine, the school newsletter, or the local newspaper.

VALUING YOUR STUDENTS' OWN IDEAS

To me, one of the most important aspects of professional development is learning to hear and value what students are saying. That may sound simple, but in reality it takes a particular talent, one that you can develop over time.

Becoming an active listener

Whenever you find yourself in a position to observe children, listen to them. *Really* listen. That is, listen to their ideas, how they say what they say, and what they think. Start now to become an active listener for the sake of your future students and other children you come in contact with.

Let's review some of the science stories in Part Two and reflect on what they tell us about the value of children's own ideas. Remember the student who wanted to see how long the icicle would *remain* an icicle while the rest of the class wanted to see how fast the icicle would melt (Chapter 4)? That story reminds me of the importance of honoring the learner as he or she tries to construct personal meaning. Other reminders of the individuality of children's thinking pervade Part Two. We saw children who wanted to:

- Weigh different liquids, thereby leading themselves to the concept of density, which was not in the original lesson plan (Chapter 9).
- Plant their own favorite foods (Chapter 8).
- Place a weird fruit in the science corner (Chapter 6).
- Investigate whether eggs change in weight when they are scrambled (Chapter 7).
- Add moons to their model of the solar system (Chapter 10).
- Explore why the moon looks bigger when it is low in the sky (Chapter 10).
- Build a lighted shoebox house and other electrical devices (Chapter 11).

Learning together with your students

When given the opportunity and the sense that their thoughts are valued, students come up with exciting ideas that enrich their way of knowing nature and the world—and enrich your own understanding as well. Learn with them and from them, and they will learn with you and from you. In the elementary school classroom, we share our scientific selves.

HOW AM I DOING? A GUIDE TO SELF-EVALUATION

Science teaching, like all other teaching, is a commitment to becoming a lifelong learner. As you reflect on what you have learned and how you teach, the checklist in Figure 17.1 may help you evaluate your professional growth. Revisit the table's questions from time to time.

Using a video to review your progress

If you are unsure of your progress in any of the areas listed in the table, ask a colleague to videotape you as you teach. Videotapes are often useful as personal guides for developing professionally. Use the table's checklist to review the videotape and assess your own progress.

But don't let the checklist—or any other teaching guidelines—intimidate you. Go easy on yourself at first. Be prepared to fumble now and again. You are traveling in what may often be uncharted territory. Do not be discouraged—science takes time.

FIGURE 17.1   **A Checklist for Self-Evaluation**

1. Am I providing *opportunities* for my students to explore natural phenomena?
   Yes _____ No _____ Comments _____
   _____

2. Do I allow my students enough *time* to complete their explorations?
   Yes _____ No _____ Comments _____
   _____

3. Whose *voice* is dominant during science explorations: mine or the children's?
   Theirs _____ Mine _____ Comments _____
   _____

4. Do I encourage children to ask *questions?* How do I do that?
   Yes _____ No _____ Comments _____
   _____

5. Do I *coach* children to find their own answers?
   Yes _____ No _____ Comments _____
   _____

6. Do I *record my students' questions* to save until another day if there is no time to help them find
   the answers when the question is asked?
   Yes _____ No _____ Comments _____
   _____

7. Do I encourage the use of *technology* for communication, presentation, and research?
   Yes _____ No _____ Comments _____
   _____

8. Do I *let go of my prepared plan* if the students' own ideas generate a new direction?
   Yes _____ No _____ Comments _____
   _____

9. Do I use *diverse techniques* to assess what the children know about a topic?
   Yes _____ No _____ Comments _____
   _____

10. Does my classroom have a *student-directed science corner?*
    Yes _____ No _____ Comments _____
    _____

11. Am I *having a good time* doing science with my students?
    Yes _____ No _____ Comments _____
    _____

LOOKING BACK TO LOOK AHEAD: A NEW CHAPTER IN YOUR SCIENCE AUTOBIOGRAPHY

Science Stories has been an attempt to have a conversation with you about science teaching and learning—about yourself as a science learner and a science teacher. As one of the goals for the book, I hoped to engage you in the stories of real children and real teachers who were learning science together and separately. To encourage your professional growth, I've also tried to connect you with some of the research that has influenced this way of doing science with children.

Some of this book's major themes

These are the major themes we have addressed:

- The nature of science
- Constructivism
- Prior knowledge
- Activity-based learning
- Science as inquiry
- Integrating technology with science
- Alternative conceptions
- Lesson planning
- Working with cooperative learning groups
- Developing curriculum
- Creating assessments
- Using community resources
- Science in your—and your students'—everyday life
- Locating your own scientific self!

SCIENCE IS TRULY FOR EVERYONE.

—Janice Koch

Science teaching needs to encourage, invite, engage, excite, interrogate, and challenge. I also like to say that it should shine like a beacon, signaling that science is truly for everyone. Science teaching can make connections to students' lived experiences and help them frame their own questions about natural phenomena as a way of making meaning of their world. I always hope that children will take ownership of their own knowledge and gain autonomy as they seek their own answers. But everything I hope for the children, I also hope for you, their science teacher.

Your next chapter

As a result of your own journey toward science teaching, you may be ready to begin the next chapter of your science autobiography. You may even want to start writing your own science stories. My wish is that your autobiography and your science journal will show a growing sense of wonder about the natural world.

Key Terms

professional development *(p. 356)*
action research *(p. 360)*

Resources for Further Exploration

Electronic Resources

Teachers.Net. **http://www.teachers.net**. A web site that offers chatboards, a reference desk, and more—many different resources to use in your teaching.

Houghton Mifflin's Project-Based Learning Site. **http://hmco.com/hmco/college/education/pbl/index.html**. Offers the chance to learn about project-based learning by doing it.

Print Resources

Darling-Hammond, L. (1998). Teacher learning that supports student learning. *Educational Leadership,* 55 (5):6–11.

Elliot, J. (1991). *Action Research for Educational Change*. Philadelphia: Open University Press.

Hopkins, D. (1993). *A Teacher's Guide to Classroom Research*. Philadelphia: Open University Press.

Kochendorfer, L. (1994). *Becoming a Reflective Teacher*. Washington, DC: National Education Association.

More Resources for Teachers

- **Organizations and Associations for Professional Development**
- **Useful Equipment, Materials, and Suppliers**
- **Safety Tips for Science in Elementary Schools**
- **Trade Books About Nature and Science**

Organizations and Associations for Professional Development

This list includes both educational and scientific organizations, as well as some groups that are specifically devoted to science teaching. By contacting the ones most suited to your own interests and needs, you can find ways to broaden your perspectives and further your own professional development as a teacher of science.

American Association for the Advancement of Science (AAAS)
1333 H Street, NW, Washington, DC 20005
(202) 326-6400

American Chemical Society (ACS)
1155 16th Street, NW, Washington, DC 20036
(202) 872-6179

American Physical Society
335 East 45th Street, New York, NY 10017
(212) 682-7341

American Zoo and Aquarium Association
7970-D Old Georgetown Road,
Bethesda, MD 20814
(301) 907-7777

Association for Supervision and Curriculum Development (ASCD)
125 North West Street, Alexandria, VA 22314-2798
(703) 549-9110

Association for Women in Science (AWIS)
2401 Virginia Avenue, NW, No. 303,
Washington, DC 20037
(202) 833-2998

Association of Science-Technology Centers
1025 Vermont Avenue, NW, Suite 500,
Washington, DC 20005-3516
(202) 783-7200

Council for Elementary Science International (CESI)
Department of Curriculum and Instruction,
212 Townsend Hall, University of Missouri,
Columbia, MO 65211
(314) 882-7247

ERIC Clearinghouse for Science, Mathematics, and Environmental Education
Ohio State University, 1200 Chambers Road, Room 310, Columbus, OH 43212-1792

ERIC Clearinghouse on Elementary and Early Childhood Education
University of Illinois, College of Education,
805 West Pennsylvania Avenue,
Urbana, IL 61801

ERIC Clearinghouse on Handicapped and Gifted Children
Council for Exceptional Children, 1920 Association Drive, Reston, VA 22091-3660
(703) 620-3600

Girls Clubs of America, Operation SMART
30 East 33rd Street, New York, NY 10016
(212) 689-3700

Harvard-Smithsonian Center for Astrophysics
Project STAR, 60 Garden Street,
Cambridge, MA 02138
(617) 495-9798

International Technology Education Association (ITEA)
1914 Association Drive, Reston, VA 22901-1502
(703) 860-2100

JASON Project
JASON Foundation for Education, 395 Totten Pond Road, Waltham, MA 02154
(617) 487-9995

Lawrence Hall of Science (LHS)
University of California, Berkeley, CA 94720
(510) 642-7771

National Assessment of Educational Progress (NAEP)
P.O. Box 6710, Princeton, NJ 08541-6710
(609) 734-1624

National Association for the Education of Young Children
1509 16th Street, NW, Washington, DC 20036-1426
(800) 424-2460

National Association for Research in Science Teaching (NARST)
c/o Dr. William Holliday, 402 Teachers College, University of Cincinnati, Cincinnati, OH 45221
(513) 475-2335

National Association for Science, Technology, and Society (NASTS)
c/o Dr. Robert Yager, University of Iowa, 765 Van Allen Hall, Iowa City, Iowa 52242-1478
(319) 335-1188

National Board for Professional Teaching Standards
300 River Place, Suite 3600, Detroit, MI 48207
(313) 259-0830

National Center for the Improvement of Science Teaching and Learning (The NETWORK, Inc.)
290 South Main Street, Andover, MA 01810
(617) 470-1080

National Center for Science Teaching and Learning (NCSTL)
Ohio State University, Research Center, 1314 Kinnear Road, Columbus, OH 43212
(614) 292-3339

National Diffusion Network (NDN)
Office of Education Research and Improvement, U.S. Department of Education, 555 New Jersey Avenue, NW, Washington, DC 20208-1525
(202) 357-6134

National Oceanic and Atmospheric Administration (NOAA)
National Weather Service Office of Warning and Forecast, 8060 13th Street, Silver Spring, MD 20910
(301) 443-8910 or (703) 243-1700

National Research Council; Center for Science, Mathematics and Engineering Education
2101 Constitution Avenue, Washington, DC 20418
(202) 334-2353

National Science Foundation (NSF)
4201 Wilson Boulevard, Alexandria, VA 22230
(703) 306-1234

National Science Resources Center (NSRC)
Arts and Industries Building, Room 1201, Smithsonian Institution, Washington, DC 20560
(202) 357-2555

National Science Teachers Association (NSTA)
1840 Wilson Boulevard, Arlington, VA 22201-3000

National S.E.E.D. Project on Inclusive Curriculum: Seeking Educational Equity and Diversity
Dr. Peggy McIntosh Center for Research on Women, Wellesley College, 106 Central Street, Wellesley, MA 02181-8259
(617) 283-2500

National Urban Coalition
1875 Connecticut Avenue, NW, Suite 400, Washington, DC 20009
(202) 986-1460 or (800) 328-6339

National Wildlife Federation
1412 16th Street, NW, Washington, DC 20036
(202) 790-4360

Native American Science Education Association (NAESA)
1333 H Street, NW, Washington, DC 20005
(202) 371-8100

Network for Portable Planetariums
c/o Ms. Sue Reynolds, Planetarium Specialist, Onondaga-Cortlandt-Madison BOCES, P.O. Box 4774, Syracuse, NY 13221
(315) 433-2671

Northwest EQUALS, FAMILY, SCIENCE
Portland State University, P.O. Box 1491, Portland, OR 97201-1491
(503) 725-3045

Project WILD
5430 Grosvenor Lane, Bethesda, MD 20814
(301) 493-5447

School Science and Mathematics Association (SSMA)
Bowling Green State University, 126 Life Sciences Building, Bowling Green, OH 43403
(419) 372-7393

Smithsonian Institution Office of Elementary and Secondary Education
Arts and Industries Building, Room 1163, Washington, DC 20560
(202) 357-2425

Society for the Advancement of Chicanos and Native Americans in Science
Thinmann Laboratories, University of California, Santa Cruz, CA 95064
(408) 429-2295

Triangle Coalition for Science and Technology Education
5112 Berwyn Road, 3rd Floor, College Park, MD 20740
(301) 220-0870

U.S. Geological Survey
Distribution Branch, Box 25286, Federal Center, Denver, CO 80225
(703) 648-6515

Young Astronaut Council
Box 65432, 1211 Connecticut Avenue, NW, Washington, DC 20036
(202) 682-1986

Useful Equipment, Materials, and Suppliers

The science stories in this book demonstrate that you don't need a great deal of fancy equipment to do science with children. Often, though, some basic supplies are extremely useful. This section lists handy materials and equipment and then the names and addresses of suppliers.

Materials and Equipment

Aluminum foil
Aluminum foil pie pans
Baby food jars
Baking soda
Balloons
Batteries: AA and D
Bell wire
Bottles: juice and soda
Buttons
Cages
Candles
Corn oil
Cornstarch
Corn syrup
Drinking straws
Flashlights
Food coloring
Funnels
Glass jars (especially large ones)
Graduated cylinders
Land snails
Magnets: all shapes and sizes
Magnifying lenses
Matches
Meter sticks
Metric measuring cups
Mirrors
Modeling clay
Owl pellets
Paper towels
Plastic bags: zip-lock style, both sandwich size and food storage size
Plastic containers: quart size or larger
Plastic cups (10 ounces, 12 ounces)
Plastic spoons
Plastic tubing
Pocket microscopes
Poster paper and paper tape

Potting soil
Prisms
Rock collections
Salt
Sand
Scales: double-pan primer balances with uniform masses
Seeds: many types, including grass, mustard, radish, red kidney beans, lima beans, pumpkin, corn
Shells
Shoeboxes
String
Sugar
Tape
Thermometers
Toothpicks
Trowels
Tuning forks
Vinegar
Wire strippers

Sources of Supplies and Equipment

Arbor Scientific
P.O. Box 2750
Ann Arbor, MI 48106-2750
(800) 367-6695

Carolina Biological Supply Company
2700 York Road
Burlington, NC 27215
(800) 334-5551

Connecticut Valley Biological Supply Co.
P.O. Box 326
Southampton, MA 01073
(800) 628-7748

Cuisenaire Co. of America
P.O. Box 5026
White Plains, NY 10602-5026
(800) 306-8050

Delta Education
P.O. Box 915
Hudson, NH 03051-0915
(800) 258-1302

Fisher Scientific Company
4901 LeMoyne Street
Chicago, IL 60651
(800) 766-7000

Nasco
901 Janesville Avenue
Fort Atkinson, WI 53538
(800) 558-9595

Radio Shack (for pocket microscopes)
Tandy Corp.
Fort Worth, TX 76102
(800) 843-7422

Science Kit and Boreal Laboratories
777 East Park Drive
Tonawanda, NY 14150
(800) 828-7777

Safety Tips for Science in Elementary Schools

The following safety suggestions are general guides for doing science in the classroom. Many of these ideas are adapted from a National Science Teachers Association booklet, *Safety in the Elementary Science Classroom*, available directly from the NSTA (see the list of Organizations and Associations for Professional Development).

Learning About Safety Policies

- Know your school's policies regarding:
  Live animals in the classroom
  Use of electrical hot plates
  Procedures in case of an accident
  Outdoor walks and field trips

- Check state and federal regulations about school safety.

- Be sure to obtain appropriate consent from parents and guardians for out-of-classroom activities and extended investigations of living organisms.

Preparing for Classroom Safety

- When stored, all equipment and materials should be labeled clearly to avoid mistakes.

- Maintain cooperative learning groups of manageable size, and make certain that the students have responsibility for returning materials to designated storage spaces.

- Be sure your classroom has a fire extinguisher and a fire blanket.

- Glassware can be dangerous. Whenever possible, use plastic instead. When you must use glassware, be sure careful procedures are followed. Have a whisk broom and dustpan handy for sweeping up broken pieces of glass.

- Use thermometers that are filled with alcohol, not mercury.

- Keep a first-aid kit in the classroom; check that it is fully supplied.

- Caution students never to taste, touch, or inhale unknown substances.

- Minimize the use of chemicals.

- Substances that are potentially harmful should be handled only by the teacher.

- Instruct students that all accidents or injuries—no matter how small—must be reported to you immediately.

- Explain to students that it is unsafe to touch their faces, mouths, eyes, and other parts of their bodies while they are working with plants, animals, or chemical substances. They must wash their hands and clean their nails after handling these materials.

- When working on an electricity unit, warn students not to experiment with the electric current in their home circuits.

- Hands should be dry when working with electrical cords, switches, or appliances.

- In general, if any materials or procedures may present a danger, explore the possible hazards carefully, and then train students in the proper methods.

Providing for Students with Special Needs

- Take steps to ensure that students with special needs or disabilities have safe and easy access to science facilities, to the extent that is advisable according to their circumstances.

- Anticipate any special needs that these students may have in emergency situations, and be sure that these requirements can be met.

Trade Books About Nature and Science

Following are some of my favorite trade books that can help you learn about science and scientists. You may want to explore them as you develop your own scientific self.

Amato, I. (1997). *Stuff: The Materials the World Is Made Of.* New York: Basic Books.

Brenneman, R. J., ed. (1984). *Fuller's Earth: A Day with Bucky and the Kids.* New York: St. Martin's Press.

Carson, R. (1956). *The Sense of Wonder.* New York: Harper & Row.

Carson, R. (1970). *Silent Spring.* Introduction by Paul R. Ehrlich. Greenwich, CT: Fawcett.

Feynman, R. P., as told to R. Leighton. (1985). *Surely You're Joking, Mr. Feynman: Adventures of Curious Character.* Edited by Edward Hutchings. New York: Norton.

Fuller, R. B. (1992). *Cosmography: A Posthumous Scenario for the Future of Humanity.* Adjuvant, Kiyoshi Kuromiya. New York: Macmillan.

Gould, S. J. (1980). *The Panda's Thumb: More Reflections in Natural History.* New York: Norton.

Gould, S. J. (1991). *Bully for Brontosaurus: Reflections in Natural History.* New York: Norton.

Hawking, S., ed. (1992). *Stephen Hawking's A Brief History of Time: A Reader's Companion.* Prepared by Gene Stone. New York: Bantam Books.

Hazen, R. M., & Trefil, J. (1991). *Science Matters: Achieving Scientific Literacy.* New York: Doubleday.

Keller, E. F. (1983). *A Feeling for the Organism: The Life and Work of Barbara McClintock.* New York: W. H. Freeman.

Negroponte, N. (1995). *Being Digital.* New York: Knopf.

Peterson, R. T. (1979). *A Field Guide to the Birds of Texas and Adjacent States.* Boston: Houghton Mifflin.

Peterson, R. T. (1980). *A Field Guide to the Birds: A Completely New Guide to All the Birds of Eastern and Central North America.* 4th ed. Boston: Houghton Mifflin.

Peterson, R. T. (1990). *A Field Guide to Western Birds.* 3rd ed. Boston: Houghton Mifflin.

Trefil, J. (1992). *Sharks Have No Bones: 1001 Things Everyone Should Know About Science.* New York: Simon & Schuster.

Glossary

action research Refers to research projects in which classroom teachers explore some area of their teaching or some aspect of the students' learning, with the goal of improving their own or a colleague's teaching. Also called *teacher research* or *classroom research.*

alternative conception An idea that is not scientifically accurate, but represents a step toward full understanding of a concept.

angiosperm A seed plant; one that produces flowers that eventually form fruits with seeds.

Animalia The scientific name for the animal kingdom. It is commonly divided into two broad groups: animals with backbones and animals without backbones.

aphelion The point in a planet's orbit when it is farthest from the sun.

arthropod An invertebrate animal with a skeleton on the outside of its body—for example, lobsters, crayfish, crabs, shrimp, spiders, and all insects.

assessment *See* Science assessment.

asteroid A tiny chunk of planet-like material that moves around the sun. Between Mars and Jupiter, there is a belt of several thousand asteroids of different sizes.

atom The smallest part of an element that retains the properties of that element. There are 92 different kinds of atoms found naturally on earth. The atoms of each element are exactly the same as one another and different from those of any other element.

authentic assessment A process of judging how well students execute a task as part of solving a problem in a larger context. Often the execution of the task involves a type of student performance. Hence, this type of assessment is also called *performance assessment.*

Benchmarks for Science Literacy A document published by the American Association for the Advancement of Science that provides guidelines for science, technology, and mathematics education. *Benchmarks* is part of a larger science education reform movement called Project 2061, named for the year of the next return of Halley's comet to earth's orbit. *Benchmarks* addresses what students should know and be able to do in science, mathematics, and technology by the end of grades 2, 5, 8, and 12.

bivalve mollusk Mollusks with two shells connected by a muscular hinge. Clams, oysters, scallops, and mussels are all bivalve mollusks.

buoyancy The lifting force that water exerts on objects. All objects appear to be lighter in water because of the buoyancy of water.

chlorophyll The substance in leaves that allows the chemical reaction of photosynthesis to take place and that gives leaves their green color.

circuit *See* Electric circuit.

classifying Sorting objects or ideas into groups on the basis of similar properties.

comparing and contrasting Discovering similarities and differences among objects or events.

compound A combination of two or more elements in a definite proportion. For example, water is H_2O—two parts hydrogen to one part oxygen.

coniferous trees Trees that do not lose their leaves in the fall, though they are inactive during the winter. The term *coniferous* (literally, "cone bearing") refers to the woody cones that these trees produce. Also called *evergreen trees.*

constant An experimental condition that remains the same throughout a scientific investigation.

constructivism A family of theories about knowledge and learning whose basic tenet is that all knowledge is constructed by synthesizing new ideas with what we have previously come to know. This means that knowledge is not passively received. Rather, knowledge is actively built up by the learner as he or she experiences the world.

convection The method by which heat energy travels in liquids and gases. When air, for example, is warmed, it expands and takes up more space. Colder air is pulled down by gravity and pushes the warmed air upward.

cooperative learning   An instructional approach in which students work together in groups to accomplish shared learning goals.

cooperative learning group   An arrangement in which a group of students, usually of mixed ability, gender, and ethnicity, work toward the common goal of promoting each other's and the group's success.

cotyledon   The part of the seed that has food stored in it for the tiny plant. Also called the *seed leaf*.

current   *See* Electric current.

curriculum   A plan of studies that includes the ways in which the instructional content is organized and presented at each grade level. *See also* Formal science curriculum; Informal science curriculum.

deciduous tree   A tree that sheds its leaves at the end of the growing season.

density   Mathematically defined as the mass of an object divided by its volume (expressed numerically in grams per cubic centimeter). Roughly, density can be conceived as how closely packed together the particles of an object are, or as the relative number of particles that can fit into a given amount of space.

design   The technological counterpart to the science-as-inquiry process. For students, the design process typically involves solving a problem by constructing a product that meets a set of established criteria.

dicot   Seed-producing plant with two seed leaves or cotyledons. Dicots include most flowers, vegetable shrubs, bean plants, and flowering trees.

discovery learning   A phrase popularized by learning theorist Jerome Bruner, who suggested that at any given stage of cognitive development, teaching should proceed in a way that allows children to discover ideas for themselves.

electric circuit   A pathway for an electric current. A circuit requires some source of electrical power, such as a battery or generator, and a material through which electrons can travel, such as copper wire. In elementary school science, circuits usually include some simple appliance that uses electricity, like a light bulb or a bell.

electric current   A flow or motion of electrons. Electric currents in wires are caused by electrons moving along the wire.

electricity   A form of energy produced by a flow of electrons.

electron   A negatively charged particle found in the atoms of all elements. Each electron is thought to be a particle of negative electricity.

element   A substance made up of only one type of material. The simplest form of matter, elements are the building blocks of all other substances. There are 92 naturally occurring elements in the universe, and 14 others that have been produced by scientists in laboratories. Iron, nickel, gold, silver, oxygen, hydrogen, helium, carbon, and mercury are all elements.

ellipse   An oval shape with no center but with two points of reference called *foci*. Planets travel in elliptical orbits around the sun, which is located at one of the foci of the orbits.

emerging relevance   A perception by students that questions or ideas arising from an investigation have personal significance to them. The key part of this concept is that certain matters become relevant to students as they engage in learning activities. By helping them explore these emerging questions and ideas, teachers can help students construct their own meaning.

energy   The ability to do work. (*Work* is defined as a force moving through a distance.)

evaluation   The process of making value judgments, based on the results of assessments, about a student's or a group's achievement in a learning area. Many educators include this process as part of the definition of assessment. *See* Science assessment.

fair test   An investigation in which all the experimental conditions remain the same (constant) except the one being tested for (the variable).

family science night   A community-involvement activity in which students bring a significant adult or adults to the school to participate in inquiry-based science activities, typically set up at stations in a large classroom or in the school cafeteria.

faults   Large cracks or breaks in the earth's crustal plates. Earthquakes are most common in areas where there are major faults.

focus   One of two points (foci) used to construct an ellipse.

focusing questions   Questions that prompt students to come up with their own ideas about a specific topic or investigation. Teachers also use focusing questions to probe their students' understandings of a particular concept related to the investigation.

formal science curriculum   The explicit statement by a school or a school district of the science topics and methodologies to be implemented at each grade level.

framework   A document, usually prepared at the state level, that offers guiding principles for elementary and secondary curricula. In science, recent state frameworks have usually been aligned with the content standards of the *National Science Education Standards.* These state frameworks tend to dictate science content, specifying the topics that should be addressed at each grade level.

fruit   A ripened ovary of a seed plant; a container for the plant's seeds.

fungi   A kingdom of living things containing tiny plantlike organisms. Fungi do not have chlorophyll and cannot make their own food. Fungi include molds, mildews, yeast, and mushrooms.

gas   A state of matter that has no definite shape or size. When a gas is poured into a container, it spreads out until it has the same size as the container and takes the shape of the container.

germination   The process by which a seed sprouts and begins to grow into a new plant.

graduated cylinder   A device used for measuring liquid volume. In essence, it is a scientific measuring cup—a glass or plastic cylinder that is calibrated in milliliters. It is a handy tool in the elementary classroom.

high tide   High level of ocean waters, when the waters reach farthest onto land. On the side of the earth facing the moon, high tide occurs because the waters bulge out from the moon's gravitational pull. At the same time, on the opposite side of the earth, there is also a high tide, because the moon's gravitational force has pulled the solid earth as well, leaving the waters bulging out on that side.

hypermedia authoring tools   Computer software packages that students can use to design and construct their own products. This type of software typically allows users to combine text, graphics, videos, and sound.

hypothesis   An inference or a guess that is tested through a planned investigation or experiment.

inference   Reasonable explanation that we construct on the basis of our observations. Inferences sometimes lead us to set up further investigations.

informal science curriculum   Science learning experiences that go beyond the formal science curriculum of the school or school district to include other topics and methodologies that connect to students' daily life outside the school. The topics might arise, for example, from spontaneous natural occurrences, local news events, or materials the students bring from home.

inquiry   The type of exploration that lies at the heart of scientific activity. According to the *National Science Education Standards,* inquiry is "a multifaceted activity that involves making observations; posing questions; . . . planning investigations; . . . using tools to gather, analyze, and interpret data; proposing answers, explanations, and predictions; and communicating the results" (National Research Council, 1996, p. 23).

instructional technology tools   Materials that aid instruction and learning. Examples include computers with Internet access, video camcorders, and authoring software such as HyperStudio and PowerPoint.

interfacing instruments   Laboratory instruments that come with a means to connect them to the personal computer. The most common such devices are probes with related software.

interview, for assessment   *See* Science interview.

interviewing a scientist   A learning activity in which students or teachers have a formal encounter with a research scientist. The interview provides the opportunity to explore the scientist's reasons for choosing a scientific career and the nature of his or her work. It is a way to change the culturally constructed "mad scientist" image that many adults and children hold.

invertebrate An animal without a backbone. Invertebrates are the most numerous of all animals.

journal *See* Science journal.

kinetic energy The energy of motion; that is, the energy that an object has because it is moving.

kingdom The broadest division of classification of living things. There are five kingdoms of living organisms: Animalia, Plantae, Protista, Fungi, and Monera.

lesson In science instruction, the process of engaging students in a meaningful science experience and in reflections on the experience.

lesson plan A document describing a teacher's plans for a particular lesson, including the ideas presented in the lesson, the activities that students will engage in, and the ways in which the teacher will help students reflect on their experiences.

liquid A state of matter with a definite size but no definite shape. A liquid takes the shape of its container.

low tide Low level of ocean waters that occurs because, as the waters bulge on two sides of the earth, the remaining waters flatten.

mass The amount of matter that is in an object, typically measured in grams and kilograms.

matter Anything that has mass and takes up space.

meaningful science experience An activity that engages students in the key processes of science, such as observing and predicting, inferring and hypothesizing, manipulating objects, investigating, and imagining; that relates to the students' everyday lived experiences; and stimulates the students to reflect on what they are exploring and come up with their own ideas.

measuring Determining distance, volume, mass, or time by using instruments that indicate these properties (e.g., centimeter sticks, graduated cylinders, scales, stopwatches).

mediator A teaching role in which the teacher helps students to learn by reflecting their own ideas back to them and guiding them in sorting out the inconsistencies. As a mediator, the teacher helps students delve deeply into their thoughts and expand their own thinking about an idea.

mixture Any combination of elements, compounds, or other mixtures. Because there are only 92 naturally occurring elements, most matter consists of either compounds or mixtures.

model A representation of a system or object: for example, a physical structure that imitates a smaller or larger structure, a mental construct that represents an object or process, or a computer program that parallels the workings of a larger system.

molecule The smallest part of a compound that still has the properties of that compound. Molecules are made up of atoms.

mollusk An invertebrate animal that has a soft, fleshy body and a protective shell made of lime—for example, clams, oysters, scallops, snails, octopuses, and squids.

monera A kingdom of living things consisting of bacteria, tiny organisms that do not have chlorophyll and cannot make their own food. All bacteria are made up of just one cell. Blue-green algae are also a part of the monera kingdom.

monocot A type of seed-producing plant with just one cotyledon or seed leaf—for example, lilies, tulips, irises, onions, and grasses and grains. Also called *monocotyledon plants*.

MST initiative A curriculum transformation project that strives to integrate mathematics, science, and technology by promoting the use of inquiry and design in problem-solving activities.

National Science Education Standards A set of standards prepared by the National Research Council and published in book form in 1996. It offers guidelines for teachers, teacher educators, curriculum developers, and school districts for establishing science education programs. The overall theme is that acquiring scientific knowledge, understanding, and abilities should be a central aspect of education, just as science has become a central aspect of our society.

national standards Guidelines written by national agencies and members of professional organizations for establishing comprehensive programs of study. Such guidelines have been published for precollege education in several discipline areas, including science, mathematics, technology, language arts, and social studies.

observation   Perceptions of an object or an event, using as many senses as possible.

open-ended questions   Questions that lead to multiple answers. They are especially important because they help students think critically about their science experiences.

open house   *See* Science open house.

orbit   The path of one heavenly body as it travels around another heavenly body.

parallel circuit   An electric circuit in which each device has a separate pathway, its own separate branch of the circuit.

performance assessment   *See* Authentic assessment.

perihelion   The point in a planet's orbit when it is closest to the sun.

phases of the moon   The various aspects of the moon as seen from earth, such as the full moon and new moon. These phases reflect changes in the amount of the lighted surface of the moon that we can see from earth.

photosynthesis   The process by which green plants use carbon dioxide from the air and water from the soil, in the presence of sunlight, to manufacture molecules of glucose. Glucose is a simple sugar that the cells of the plant then use to make energy for the plant to carry on its functions, including growth.

physical science   The branch of elementary school science that includes the exploration of nonliving materials, their interactions, and interactions between matter and energy.

planets   Principal members of the solar system that move around the sun. In order of their distance from the sun, the planets are Mercury, Venus, Earth, Mars, Jupiter, Saturn, Uranus, Neptune, and Pluto. Planets shine by reflecting the light of the sun or other stars.

planning an investigation   Determining a reasonable procedure that could be followed to test an idea. This includes listing the materials needed, writing out the procedure to be followed, and identifying which variables will be kept the same and which will be changed.

portfolio   *See* Science portfolio.

potential energy   The energy an object has because of its position; stored-up energy. When potential energy is set free, it is changed into kinetic energy.

predicting   Estimating the outcome of an event on the basis of observations and, usually, prior knowledge of similar events.

prior knowledge   What an individual has learned from all his or her previous experiences. This plays a crucial role in determining how the person integrates a new concept.

process skills   Abilities that help people gain information about nature and natural phenomena: observing, inferring, classifying, recording data, predicting, and planning investigations. These skills are employed on a planned and regular basis by those engaged in scientific activity. Also called *inquiry skills*.

professional development   The process by which teachers strive to improve their work as teachers in order to grow in their profession. It is generally based on inquiry into their own teaching practices, active engagement in their own research, and teacher workshops and courses.

property words   The basic words we use to describe the material world, referring to common properties of objects such as size, shape, color, odor, texture, taste, composition, and hardness.

protista   A kingdom of simple organisms, consisting mainly of algae, except for blue-green algae. Other protists are protozoans like ameba, paramecium, and euglena.

questions   *See* Focusing questions; Open-ended questions.

recording data   Writing down (in words, pictures, graphs, or numbers) the results of observations of an object or an event.

reflective teacher   A teacher who thinks deeply about his or her teaching practices, the needs and identities of the students, and what the teaching is intended to accomplish.

rubric   A set of criteria used to determine the scoring value of an assessment task.

satellite   Any heavenly body that travels around another heavenly body. The moon is a satellite of the earth, and the earth-moon system is a satellite of the sun.

science assessment A process of collecting information that is used to determine the quality and character of an individual or group performance in a science learning experience.

science autobiography A personal description of one's experience with science, in or out of school.

science circus A science activity that consists of several stations at which the visitors are asked to perform certain tasks and record their results or reactions.

science conversation (or science interview) A direct discussion between a teacher and an individual student about a science learning experience. This activity allows teachers to ask students to elaborate on their ideas—a good way to determine the depth of a student's understanding. Through such conversations, teachers gain important clues about how students construct personal meaning, and these clues then influence future instruction.

science interview A conversation between student and teacher that is designed to assess concept understanding.

science journal A personal journal in which the writer focuses on nature and natural events in his or her daily experiences.

science open house A community event in which students demonstrate one or more science experiments, activities, or models. This event, usually presented by children in the upper elementary grades (grades 4–6), typically involves several groups of students, and it is set up in different classrooms in the evening.

science portfolio A selection of a student's work during a semester or a unit. The work in the portfolio might include one or more reports, drawings, poems, or other products representing what the student has learned in science over a defined period of time. The most important feature of a portfolio is that the student selects its contents and reflects on his or her reasons for this selection.

scientific literacy The ability to use the processes of science to make important life decisions; in particular, the ability to explore a problem through careful reasoning.

seed The part of the plant that can grow into a new plant.

seed plant *See* Angiosperm.

series circuit An electric circuit in which the current has only a single pathway through the entire circuit.

solar system The sun and the group of heavenly bodies that move around it. The sun is the only member of our solar system that is a star.

solid A state of matter with a definite size and shape.

spiraling of curriculum Engaging students in the same topic of study at different grade levels, so that the topic is explored at greater depth in later grades.

star A heavenly body that produces its own light. There are billions of stars in the universe, including our sun. About 3,000 stars are visible with the naked eye.

states of matter The basic forms in which matter is found: solid, liquid, and gas (and a state that is rare on earth, plasma).

station assessment An assessment activity in which work stations are set up around the room to create a genuine context for the students to perform tasks. The students typically use an answer sheet to respond to the questions at each station. Each work station assesses a different aspect of the science unit.

suspension A mixture in which solid particles do not dissolve, but literally are suspended in a liquid.

sustained inquiry Prolonged investigation of a scientific phenomenon. For elementary school science, this typically means an investigation that includes multiple activities over the course of several days or weeks.

taxonomy An entire classification system, such as the scientific classification of all living things into categories ranging from kingdom to species.

testing Assessment of students' learning by means of teacher-made tests or state and local tests designed by educational agencies or testing services.

theory An idea that has been tested and (to some significant degree) proved. Even though they have been "proved," scientific theories are often in a state of revision as further evidence accumulates.

tide   *See* High tide; Low tide.

unit of study   A segment of the curriculum that includes several lessons designed around a central theme or topic.

univalve mollusk   Mollusks with one shell, usually shaped in a spiral. Snails, slugs, and conches are univalve mollusks.

variable   The property or condition of a scientific investigation that will change when experimental conditions are changed.

vegetable   An edible plant part that is the root, stem, leaf, or flower of the plant.

vertebrate   An animal with a backbone. There are five vertebrate groups: fish, amphibians, birds, reptiles, and mammals.

volume   The amount of space an object takes up.

Liquid volume is typically measured in milliliters and liters; solid volume, in cubic centimeters and cubic meters. When we refer to the *size* of an object, we usually mean its volume.

wait time   The time that elapses between the moment a teacher asks a question and the moment the teacher selects a student to respond, offers a clue, rephrases the question, or otherwise moves ahead with the lesson.

weathering   The wearing away of rocks on the earth by the action of the sun, air, and water.

weight   The gravitational pull that the earth has on an object. The weight of an object increases when its mass increases; but weight in scientific terms is not the same as mass, because weight is dependent on the gravitational pull that is exerted on the object.

year   The time needed for a planet to make one complete turn or revolution about the sun.

References

American Association for the Advancement of Science. (1993). *Benchmarks for Science Literacy.* Washington, DC: Author.

Abder, P. (1990). Elementary science process circus. Paper presented at the National Science Teachers Association Regional Meeting, San Juan, Puerto Rico.

Barba, R. H. (1998). *Science in the Multicultural Classroom: A Guide to Teaching and Learning.* (2nd ed.) Needham Heights, MA: Allyn and Bacon.

Barrow, L. H., Knipping, N., & Litherland, R. (1996). Evening science: Solving science problems. *Science and Children,* 34(2): 20–30.

Barton, A. C. (1998). Teaching science with homeless children: Pedagogy, representation and identity. *Journal of Research in Science Teaching,* 35(4): 379–394.

Brooks, J. G., & Brooks, M. (1993). *In Search of Understanding: The Case for Constructivist Classrooms.* Alexandria, VA: Association for Supervision and Curriculum Development.

Bruner, J. S. (1960). *The Process of Education.* Cambridge, MA: Harvard University Press.

Bruner, J. S. (1966). *Toward a Theory of Instruction.* Cambridge, MA: Harvard University Press.

Chancer, J., & Rester-Zodrow, G. (1998). *Moon Journals: Writing, Art and Inquiry.* Portsmouth, NH: Heinemann.

Chuska, K. R. (1995). *Improving Classroom Questions.* Bloomington, IN: Phi Delta Kappa.

Clewell, B. C., Anderson, B. T., & Thorpe, M. E. (1992). *Breaking the Barriers: Helping Female and Minority Students Succeed in Mathematics and Science.* San Francisco: Jossey-Bass.

Cobb, V. (1972). *Science Experiments You Can Eat.* Philadelphia: Lippincott.

Conant, J. B. (1966). *Science and Common Sense.* New Haven: Yale University Press.

Dana, T. M., Lorsbach, A. W., Hook, K., & Briscoe, C. (1991). Students showing what they know: A look at alternative assessments. In G. Kulm & S. M. Malcom (Eds.), *Science Assessment in Service to Reform.* Washington, DC: American Association for the Advancement of Science.

Darling-Hammond, L. (1998). Teacher learning that supports student learning. *Educational Leadership,* 55(5):6–11.

Dewey, J. (1904). The relation of theory to practice in education. In C. McMurray (Ed.), *The Relation of Theory to Practice in the Education of Teachers: Third Yearbook for the National Society of the Scientific Study of Education.* Chicago: University of Chicago Press.

Dewey, J. (1933, 1988). *How We Think.* Boston: Houghton Mifflin.

Doran, R., Lawrenz, F., & Helgeson, S. (1994). Research on assessment in science. In D. Gabel (Ed.), *Handbook of Research on Science Teaching and Learning.* A project of the National Science Teachers Association. New York: Macmillan.

Driver, R. (1989). Students' conceptions and the learning of science. *International Journal of Science Education,* 11:481–490.

Dublin, P., Pressman, H., & Barnett, E. (1994). *Integrating Computers in Your Classroom: Elementary Science.* New York: HarperCollins.

Duckworth, E. (1991). Twenty-four, forty-two, and I love you: Keeping it complex. *Harvard Educational Review,* 61(1):1–24.

Duckworth, E. (1996). *The Having of Wonderful Ideas and Other essays on Teaching and Learning.* New York: Teachers College Press.

Duckworth, E., Easley, J., Hawkins, D., & Henriques, A. (1990). *Science Education: A Minds-on Approach for the Elementary Years.* Hillsdale, NJ: Erlbaum.

Elliot, J. (1991). *Action Research for Educational Change.* Philadelphia: Open University Press.

Eltgeest, J. (1985). The right question at the right time. In W. Harlen (Ed.). *Primary Science: Taking the Plunge.* Portsmouth, NH: Heinemann.

Feynman, R. P. (1968). What is science? *Physics Teacher,* 7(6):313–320.

Fleischmann, P. (1988). *Joyful Noise: Poems for Two Voices.* New York: Harper & Row.

Fort, D., & Varney, H. (1989). How students see scientists: Mostly male, mostly white and mostly benevolent. *Science and Children,* 26:8–13.

Gage, N. L., & Berliner, D. C. (1998). *Educational Psychology.* 6th ed. Boston: Houghton Mifflin.

Gallas, K. (1995). *Talking Their Way into Science.* New York: Teachers College Press.

Gilster, P. (1997). *Digital Literacy.* New York: Wiley.

Goldhaber, J. (1994, Fall). If we call it science, can we let the children play? *Childhood Education,* pp. 24–27.

Gould, S. J. (1981). *Hen's Teeth and Horse's Toes.* New York: Norton.

Grabe, M., & Grabe, C. (1998). *Integrating Technology for Meaningful Learning.* Boston: Houghton Mifflin.

Grumet, M. (1980). Autobiography and reconceptualization. *Journal of Curriculum Theorizing,* 2(2):155–158.

Grumet, M. (1991). The politics of personal knowledge. In C. Witherell & N. Noddings (Eds.), *Stories Lives Tell: Narratives and Dialogue in Education.* New York: Teachers College Press.

Hawkins, D. (1965). Messing about in science. *Science and Children,* 2(5).

Hopkins, D. (1993). *A Teacher's Guide to Classroom Research.* Philadelphia: Open University Press.

Hubbard, R., & Wald, E. (1993). *Exploding the Gene Myth.* Boston: Beacon Press.

Johnson, D., & Johnson, R. (1994). *Learning Together and Alone.* Boston: Allyn and Bacon.

Jones, C., & Levin, J. (1994). Primary/elementary teachers' attitudes toward science in four areas related to gender differences in students' science performance. *Journal of Elementary Science Education,* 6(1):46—65.

Kahle, J. B., & Meece, J. (1994). Research on girls and science: Lessons and applications. In D. Gabel (Ed.), *Handbook of Research in Science Teaching and Learning.* Washington, DC: National Science Teachers Association.

Kaner, E. (1989). *Balloon Science.* Reading, MA: Addison-Wesley.

Koballa, T. R., & Crawley, F. E. (1985). The influence of attitude on science teaching and learning. *School Science and Mathematics,* 85(3): 222–232.

Kober, N. (1993). *Edtalk: What We Know About Science Teaching and Learning.* Washington, DC: Council for Educational Development and Research.

Koch, J. (1990). The science autobiography project. *Science and Children,* 28(3):42–44.

Koch, J. (1993a). Face to face with science misconceptions. *Science and Children,* 31(3):39–41.

Koch, J. (1993b). *Lab Coats and Little Girls: The Science Experiences of Women Majoring in Biology and Education at a Private University.* Ann Arbor, MI: University Microfilms International No. 5712.

Kochendorfer, L. (1994). *Becoming a Reflective Teacher.* Washington, DC: National Education Association.

Lederman, N. G. (1992). Students' and teachers' conceptions of the nature of science: A review of the research. *Journal of Research in Science Teaching,* 29(4):331—359.

Logan, J. (1997). *Teaching Stories.* New York: Kodansha International.

Loucks-Horsley, S. (1990). *Elementary School Science for the 90s.* Andover, MA: The Network, Inc.

Margulis, L., & Schwartz, K. V. (1982). *Five Kingdoms: An Illustrated Guide to Phyla of Life on Earth.* New York: Freeman.

Mark, J. (1992, June). *Beyond Equal Access: Gender Equity in Learning with Computers.* Newton, MA: Women's Educational Equity Act Publishing Center.

McIntosh, M. (1983). *Interactive Phases of Curricular Re-Vision: A Feminist Perspective.* (Working Paper No. 124.) Wellesley, MA: Wellesley College Center for Research on Women.

National Commission on Excellence in Education. (1983). *A Nation at Risk.* Washington, DC: U.S. Government Printing Office.

National Research Council. (1996). *The National Science Education Standards.* Washington, DC: National Academy Press.

National Science Board. (1983). *Educating Americans for the 21st Century.* Washington, DC: National Science Foundation.

Nelson, C., & Watson, J. A. (1991). The computer gender gap: Children's attitudes, performance, and socialization. *Journal of Educational Technology Systems,* 19(4):345–353.

Noddings, N. (1990). Constructivism in mathematics education. *Journal for Research in Mathematics Education,* no. 4.

Nussbaum, J., & Novak, J. D. (1976). An assessment of children's concepts of the earth utilizing structured interviews. *Science Education,* 60:535–550.

Perkes, V. A. (1975). Relationship between a teacher's background to sensed adequacy to teach elementary science. *Journal of Research in Science Teaching,* 12(1):85–88.

Piaget, J. (1964). *The Construction of Reality in the Child.* New York: Basic Books.

Piaget, J. (1974). *To Understand Is to Invent: The Future of Education.* New York: Grossman.

President's Committee of Advisors on Science and Technology (1997). *Report to the President on the Use of Technology to Strengthen K–12 Education in the United States.* Washington, DC: Executive Office of the President of the United States.

Pyramid Film & Video. (1988). *A Private Universe: An Insightful Lesson on How We Learn.* Santa Monica, CA: Author.

Reynolds, K., & Barba, R. (1996). *Technology for the Teaching and Learning of Science.* Needham Heights, MA: Allyn and Bacon.

Rowe, M. B. (1974). Wait-time and rewards as instructional variables: Their influence on language, logic and fate control: Part One— Wait-time. *Journal of Research in Science Teaching,* 11(2):81–94.

Rowe, M. B. (1987). Wait-time: Slowing down may be a way of speeding up. *American Educator,* 11(1):38–47.

Rutherford, F. J., & Ahlgren, A. (1990). *Science for All Americans.* New York: Oxford University Press.

Sadker, M., & Sadker, D. (1994). *Failing at Fairness: How America's Schools Cheat Girls.* New York: Charles Scribner's Sons.

Salvia, J., & Ysseldyke, J. E. (1998). *Assessment,* 7th ed. Boston: Houghton Mifflin, 1998.

Sanders, J., Koch, J., & Urso, J. (1997). *Gender Equity Right from the Start: Instructional Activities for Teacher Educators in Mathematics, Science and Technology.* Hillsdale, NJ: Erlbaum.

Schon, D. A. (1983). *The Reflective Practitioner: How Professionals Think in Action.* New York: Basic Books.

Schon, D. A. (1986). *Educating the Reflective Practitioner.* San Francisco: Jossey-Bass.

Schon, D. A. (1991) *The Reflective Practitioner.* San Francisco: Jossey-Bass.

Shepardson, D. P., & Britsch, S. (1997). Children's science journals. *Science and Children,* 35(2): 13ff.

Shrigley, R. (1983). The attitude concept and science teaching. *Science Education,* 67(2):425–442.

Shrigley, R. (1990). Attitude and behavior are correlates. *Journal of Research in Science Teaching,* 27(2):97–113.

Siegler, R. S. (1991). *Children's Thinking.* Englewood Cliffs, NJ: Prentice-Hall.

Slavin, R. E. (1987). Cooperative learning and the cooperative school. *Educational Leadership,* 45(3):7–13.

Smith, K., & Welliver, P. (1990). The development of a science process assessment for fourth-grade students. *Journal of Research in Science Teaching,* 27(8):727– 738.

Songer, N. B., & Linn, M. C. (1991). How do students' views of science influence knowledge integration? *Journal of Research in Science Teaching,* 28(9):761–784.

Tobin, K. (1986). Effects of teacher wait time on discourse characteristics in mathematics and language arts classes. *American Educational Research Journal,* 23:191–200.

Valenza, J. K. (1997, August 21). Teachers can mine net for lesson plans, other kinds of help. *Philadelphia Inquirer.*

VanCleave, J. P. (1993). *200 Gooey, Slippery, Slimy, Weird and Fun Experiments.* New York: Wiley.

von Glasersfeld, E. (1995). *Radical Constructivism: A Way of Knowing and Learning.* London: Falmer Press.

Vygotsky, L. (1962). *Thought and Language.* Cambridge, MA: MIT Press

Walberg, H. J. (1969). Social environment as a mediator of classroom learning. *Journal of Educational Psychology,* 60:443–448.

Whitney, D. (1995). The case of the misplaced planets. *Science and Children,* 32(5):12–14ff.

Yager, R. (1991). The constructivist learning model: Towards real reform in science education. *Science Teacher,* 58(6):52–56.

Index

Abder, P., 88
Action research, 360
Active learning, 6, 7, 57
Active listening, 360
Activities
  community, 339–349.
    See also Commu-
    nity activities
  hands-on, 10–11
  to illustrate emerging
    relevance, 203–206
  planning, 255
  selecting, 307
Activity guides, 307, 309
Ahlgren, A., 296
Air, 140
Alternative conceptions
  about energy, 234
  dealing with, 63
  explanation of, 13, 63
  function of, 84–85
  misconceptions as, 84
  strategies to change
    persistent, 85
Amato, Ivan, 137
American Association for
  the Advancement
  of Science,
  295–296
Anderson, B. T., 108
Anemometers, 338, 340
Angiosperm, 166
Animalia, 166
Aphelion, 223
Arthropods, 182
Assessment
  elements of good,
    334–335
  instructional context
    and, 313
  of multiple types of
    performance,
    326–332
  National Science Edu-
    cation Standards
    and, 332–334
  nature of perfor-
    mance, 311–312
  science, 311

of student learning,
  259
using science conversa-
  tions for, 323–324
using science journals
  for, 313–321
using science portfo-
  lios for, 321–323
using technology for,
  324–325
Association for the Education of Teachers of
  Science, 289
Association for Women in
  Science, 344
Associations. *See* Professional associa-
  tions; *specific*
  associations
Asteroids, 218
Astronomical units (AUs),
  214, 215
Atoms, 142
Authentic assessment,
  311–312. *See also*
  Performance as-
  sessment

Barba, Robertta H., 3, 268,
  282
Bartholemew and the Oobleck
  (Dr. Seuss), 139
Barton, A. C., 18
Bascom, Florence, 91
Batteries, Bulbs, and Wires
  Revisited (Science
  Story), 237–246
Batteries, Bulbs, and Wires
  (Science Story),
  235–237
Benchmarks, 13–14
*Benchmarks for Science Liter-
  acy* (American As-
  sociation for the
  Advancement of
  Science), 13–14,
  63, 295, 296
Berliner, D. C., 65
Bias, modifying, 109

Bird Story (Science Jour-
  nal), 45–49
Bivalve mollusks, 182
A Book of Snails (Science
  Story), 178,
  180–181
Borun, M., 348
The Bottle and the Balloon
  (Science Story),
  54–56
Brainstorming, 92, 96
Brooks, Jacqueline Gren-
  non, 10, 57, 60,
  202
Brooks, Martin, 10, 57, 60,
  202
Bruner, Jerome, 6, 8–10, 306
Buoyancy, 199

Canal Races, 342
Cans of Soda (Science
  Story), 196–197
Certification, teacher, 359
Chlorophyll, 167
Chuska, K. R., 265
The Circus Comes to
  Mount Holly (Sci-
  ence Story), 89–99
Classifying
  explanation of, 100,
    130
  skills for, 131, 132
Clewell, B. C., 108
Cobb, Vicki, 18
Collaboration. *See also* Co-
  operative learning
  between professional
    scientists, 269
  between students and
    teachers, 184–185
  using electronic tele-
    communication,
    20–21, 283–284
Collections. *See* Nature col-
  lections
Communication, one-way
  vs. two-way, 62
Community activities
  conversations with

professional sci-
  entists as, 344–346
science nights as,
  339–344
science open houses
  as, 346–347
science resources in
  community and,
  347–349
searching out, 354
Community Connections
  Project, 348
Compare and contrast
  exercise, 101
Compounds, 142
Computers
  for creating lesson
    plans, 272–273
  effects on students of,
    19
  models generated by,
    210–211
  student access to,
    21–22
  ways of using, 19–21
Computers in the Schools,
  285
Computer software
  educational uses for,
    275–276
  evaluating, 285–286
  hypermedia authoring
    tools as, 325
  for interactive prob-
    lem-solving, 20,
    284–285
  probes with, 283–284
  simulation, 20, 285
Conant, James B., 3
Concrete experiences. *See*
  also Hands-on ac-
  tivities
  Dewey and, 64–65
  explanation of, 10–11
Condensation, 125
Coniferous trees, 126
Constants, 81–82
Constructivism
  Bruner and, 8–9

The text box at top left reads:

> *Science Stories* are central to this book's approach—showing science content, learning, and strategies in action.

Icicles

It is an icy-cold winter morning in the Northeast. It snowed two days ago, and the temperature has plummeted to well below the freezing point of water. Ice and snow cover everything.

On his way to school in this urban community, Mr. Wilson notices icicles hanging from the edges of roof lines. The icicles glisten in the sun. They are of varying lengths and thicknesses. He reaches up and breaks off some extra-long ones and brings them to his third-grade class.

The students arrive bundled up with scarves, gloves, and hats in addition to their heavy winter coats. They settle into their seats, and after the morning business, Mr. Wilson reaches to the ledge outside the classroom window, where he has been storing the icicles.

He shows them to the children. "Where do you think I found these?" he asks.

Mr. Wilson notices . . .

The text box at left reads:

> *Science Stories* **pedagogy includes:**

The text box at left reads:

> Special **"Thinking"** segments that focus on either teacher's or author's goals.

Mr. Wilson's thinking: Although this is an impromptu lesson, Mr. Wilson has an objective in mind. He's hoping the students will begin to learn what melting really is. The question about an icicle's weight before and after melting was an important one, so he stimulates their attention to it. He also guides them in framing their investigations so that they will be able to explore the changes that occur when the icicle melts. He points things out and inquires about their plans. He is somewhat like a tour guide or a coach. Unlike a tour guide, however, he does not explain the details of all the sights. Instead, he listens to the students' impressions and asks for whatever meanings they may construct.

Being a coach

The text box reads:

> **"Expanding Meaning"** sections at the end of each **Story** that provide:
> - teaching ideas behind the Story
> - science ideas behind the Story
> - questions and resources for further exploration.

EXPANDING MEANINGS

The Teaching Ideas Behind This Story

■ The way Mr. Wilson used the icy weather to engage the students in a science activity reflects an important connection between the outside environment and the activity within the classroom. Making this connection is a very important part of doing science with children. Nature is all around us, and frequently it presents itself through changes in the weather. These activities may be thought of as *informal science learning experiences* as contrasted with the more formal, prescribed science curricula that we will explore later in this book.

The Science Ideas Behind This Story

■ The activity in this story addresses what happens when matter *changes state*, in this case from a solid to a liquid. The students observe that when a certain amount of matter changes state, its mass does not change. Changes in state are associated with different amounts of energy, but the amount of mass remains constant before and after the change of state. Each group of children added heat energy to its icicle so that it would melt. The icicle absorbed the heat energy. That caused its particles to move faster and spread farther apart, turning the solid into a liquid. But the additional energy did not affect the mass.

Questions for Further Exploration

■ In what ways do you think the icicle activity stimulated the students' thinking process?

■ What prior experiences do you think the children were relying on to construct new meanings?

■ What science ideas did the teacher need to know before engaging children in this activity?

■ What materials did the teacher rely on to design this activity? What other materials might have been useful?

Resources for Further Exploration

Electronic Resources

Constructivist Classrooms. **http://129.7.160.115/INST5931/constructivist. html.** A summary and discussion of a book by J. G. Brooks and M. G. Brooks, *In Search of Understanding: The Case for Constructivist Classrooms* (Alexandria, VA: Association for Supervision and Curriculum Development, 1993). Includes a list of pointers on constructivist strategies.

Just Think: Problem Solving Through Inquiry. Videotape series available from the New York State Education Department, Office of Educational Television and Public Broadcasting, Cultural Education Center, Room 10A75, Albany, NY 12230. Offers an excellent view of teachers as mediators in real classrooms.